Public, Animal, and Environmental Aquaculture Health Issues

Public, Animal, and Environmental Aquaculture Health Issues

Edited by

Michael L. Jahncke
E. Spencer Garrett
Alan Reilly
Roy E. Martin
Émille Cole

WILEY-INTERSCIENCE
A John Wiley & Sons, Inc., Publication

This book is printed on acid-free paper. ♾

Published simultaneously in Canada.

For ordering and customer service, call 1-800-CALL-WILEY.

Library of Congress Cataloging-in-Publication Data:

ISBN: 0-471-38772-X

Printed in the United States of America.

10 9 8 7 6 5 4 3 2 1

Contents

List of Tables

Preface

Sustainable aquaculture—can it feed the world? It is unlikely, but sustainable aquaculture can successfully provide safe and wholesome aquaculture products to supplement the declining supply of wild fishery products. The Food and Agriculture Organization (FAO) defines aquaculture as "the farming of aquatic organisms, including fish, molluscs, crustaceans, and aquatic plants." Aquaculture is a rapidly growing sector of agriculture, with numerous business entrepreneurs looking to begin new aquaculture companies worldwide. Local, regional, and national governments are encouraging the development of community-friendly, sustainable aquaculture business enterprises.

Most definitions of sustainability address the need for economically viable agriculture practices that meet human food needs and are environmentally and socially positive. The World Commission on Environment and Development (WCED) defines sustainable development as "development that meets the needs of the present generation without compromising the ability of future generations to meet their own needs." The United States Department of Agriculture (USDA) has a broader definition that includes "an integrated system of plant and animal production practices having a site-specific application that will over the long term: (1) satisfy human food and fiber needs; (2) enhance the environmental quality and natural resource base upon which the agricultural economy depends; (3) make the most efficient use of non-renewable resources and on-farm resources and integrate, where appropriate, natural biological cycles and controls; (4) sustain the economic viability of farm operations; and (5) enhance the quality of life for farmers and society as a whole."

As wild fishery stocks decline, there will be increasing demands for aquaculture products. Along with this increased demand are requirements by consumers and governments to maintain a safe food supply and a healthy environment. A holistic approach is needed that uses multidisciplinary central planning efforts to anticipate and manage current and emerging shrouded public, animal, and environmental health issues associated with aquaculture. "Holistic" defines the functional relation between parts and wholes. Included in any holistic approach for aquaculture is the need to understand the intrinsic components associated with the international trade of fishery products and the potential for use of Hazard Analysis Critical Control Point (HACCP) principles as an aquaculture risk management tool.

We are grateful to the authors and reviewers who provided their contributions to this book. We are particularly indebted to Mrs. Barbara Comstock, for her expert assistance in the preparation of this book.

Michael L. Jahncke
E. Spencer Garrett
Alan Reilly
Roy E. Martin
Emille Cole

Acronyms

ADI	Allowable daily intake
APEC	Asia-Pacific Economic Cooperation
ASEAN	Countries of Brunei, Indonesia, Malaysia, Philippines, Singapore, Thailand, and Vietnam
ASP	Amnesic shellfish poisoning
BKD	Bacterial kidney disease
BMPs	Best management practices
CAC	Codex Alimentarius Commission
CCFICS	Codex Committee on Food Import and Export Inspection and Certification Systems
CCFL	Codex Committee on Food Labeling
CCVD	Channel catfish virus disease
CCP	Critical control points
Codex	Codex Alimentarius International Food Standards Programme
COFI	Committee on Fisheries
DDE	Dichlorodiphenyldichloroethrlene
DDT	Dichlorodiphenyltrichloroethane
DNA	Deoxyribonucleic acid
DSP	Diarrhetic shellfish poisoning
EC	European Community
ECS	Ecological Consultancy Services, Ltd.
EECO	European Economic Cooperation Organization
EIBS	Erythrocytic inclusion body syndrome
EMS	Environmental monitoring system

ESC	Enteric septicemia of catfish
EU	European Union
FAO	Food and Agriculture Organization of the United Nations
FBT	Fishborne trematodiasis
FCR	Feed conversion rate
FEAP	Federation of European Aquaculture Producers
GAPs	Good agricultural practices
GATS	General Agreement on Trade in Services
GATT	General Agreement in Tariffs and Trade
GESAMP	Group of Experts on the Scientific Aspects of Marine Pollution
GMF	Genetically modified foods
GMO	Genetically modified organisms
GMPs	Good manufacturing practices
ha	Hectares
HACCP	Hazard Analysis Critical Control Point
HHP	High hydrostatic pressure
HSB	Hybrid striped bass
ICES	International Council for the Exploration of the Sea
ICPM	Interim Commission on Phytosanitary Measures
IEC	International Electrotechnical Commission
IFT	Institute of Food Technologists
IHHNV	Infectious hypodermal and hematopoietic necrosis virus
IP	Identity preservation
IPN	Infectious pancreatic necrosis
IPPC	International Plant Protection Commission
ISA	Infectious salmon anemia
ISPM	International Standards for Phytosanitary Measures
ISO	International Standards Organization
ITO	International Trade Organization
JSA	Joint Subcommittee on Aquaculture
LDC	Least-Developed Countries
LIFDC	Low-Income Food Deficit Countries
MERCOSUR	Southern Common Market
MFN	Most Favored Nation
MMPA	Marine Mammal Protection Act
mt	Metric tons
NACA	Network Aquaculture Centres in Asia
NAFTA	North American Free Trade Agreement
NFI	National Fisheries Institute
NGO	Nongovernmental Organizations

NPPO	National Plant Protection Organization
NSP	Neurotoxic shellfish poisoning
OECD	Organization for Economic Cooperation and Development
OIE	Office International des Epizootics
PCB	Polychlorinated biphenyls
PCDD	Polychlorinated dibenzo-*p*-dioxins
PCDF	Polychlorinated dibenzo-*p*-difurans
PSP	Paralytic shellfish poisoning
QMP	Quality Management Program
RAS	Recirculating aquaculture systems
RNA	Ribonucleic acid
rDNA	Recombinant deoxyribonucleic acid
RPPO	Regional Plant Protection Organization
RSIVD	Red Sea bream iridoviral disease
SERNAPESCA	National Fisheries Service (Chile)
SOPs	Standard operating procedures
SPF	Specific pathogen free
SPS	Sanitary and phytosanitary
TBT	Technical Barriers to Trade
TDE	Diphenylethanedichlorophenyl (ethane)
TED	Turtle excluder device
TRIPS	Agreement on Trade-Related Aspects of Intellectual Property Rights
TSV	Taura syndrome virus
USA	United States of America
USCDC	US Centers for Disease Control and Prevention
USDA	US Department of Agriculture
USDOC	US Department of Commerce
USEPA	US Environmental Protection Agency
USFDA	US Food and Drug Administration
USNMFS	US National Marine Fisheries Service
USSR	United Soviet Socialist Republics
VDV	Viral deformity virus
VEH	Viral epidermal hyperplasia
VHS	Viral hemorrhagic septicemia
VNN	Virus nervous necrosis
WGMAFC	Working Group on Marine Fish Culture
WHO	World Health Organization
WTO	World Trade Organization
YAV	Yellowtail ascites virus
YHV	Yellowtail virus

Contributors

MALCOLM BEVERIDGE, Institute of Aquaculture, University of Stirling, Stirling FK9 4LA, United Kingdom

STUART BUNTING, Institute of Aquaculture, University of Stirling, Stirling FK9 4LA, United Kingdom

E. SPENCER GARRETT, National Marine Fisheries Service, National Seafood Inspection Laboratory, 3209 Frederic Street, Pascagoula, MS 39567, United States of America

PETER HOWGATE, 26 Lavender Row, Stedham, Midhurst, West Sussex GU29 0NS, United Kingdom

MICHAEL L. JAHNCKE, Virginia Seafood Agricultural Research and Extension Center, Virginia Tech, 102 S. King Street, Hampton, VA 23669, United States of America

CARLOS A. LIMA DOS SANTOS, International Consultant (Food Safety and Quality), Rio de Janeiro, Brazil

ROY E. MARTIN, Science Advisor, National Fisheries Institute, 11283 Hickory Ridge Court, Spring Hill, FL 34609, United States of America

ALAN REILLY, Food Safety Authority of Ireland, Abbey Court, Lower Abbey Street, Dublin 1, Ireland

MICHAEL H. SCHWARZ, Virginia Seafood Agricultural Research and Extension Center, Virginia Tech, Hampton, VA 23669, United States of America

1

Status of World Fisheries and the Role of Aquaculture

Roy E. Martin

INTRODUCTION

This book identifies issues that need to be considered for the successful development of sustainable aquaculture operations. Chapters 2 and 3 address public, animal, and environmental health issues from nonindustrialized and industrialized country perspectives. Chapter 4 discusses hazard analysis critical control point (HACCP) principles in the context of providing a risk management structure to address these issues. Chapter 5 addresses the importance of aquaculture to world trade, specifically Codex Alimentarius (Codex), General Agreements on Tariffs and Trade (GATT), and the World Trade Organization (WTO). Chapter 6 explores future needs of aquaculture and emerging issues such as the continued use of fish meal in formulated diets and the use of genetically modified organisms (GMO) in aquaculture.

Public, Animal, and Environmental Aquaculture Health Issues,
Edited by Michael L. Jahncke, E. Spencer Garrett, Alan Reilly,
Roy E. Martin, and Emille Cole.
ISBN 0-471-38772-X (cloth)

WILD STOCK STATUS

Except for the southeast Atlantic, southwest Pacific, and west central Pacific Oceans, all of the world's major fishing areas have had minor changes or a decline in landings (FAO 2000b). Thus future increases of harvests from these areas are not expected. On a global basis, wild capture fisheries production ranges from 89.5 to 93.8 million metric tons (mt), which includes tonnages for production of fish meal and oil. Tonnages of species that are harvested for fish meal, i.e., menhaden, anchoveta, capelin, etc., vary from year to year because of biological cycling. Globally, wild-captured marine species account for 90% of this production, with the remainder being harvested from inland waters (FAO 2000b).

In a recent study published in the November 2000 issue of *Fisheries Magazine*, it was stated that "82 fish species are at risk of extinction." Many of these species are commercially harvested. The list included several species of shark, skate, sturgeon, smelt, cod, sea horse, pipefish, rockfish, snook, grouper, goby, Atlantic salmon, and Atlantic halibut. Pacific cod, herring, and groundfish were listed as species at potential risk along the west coast of the United States (Ess 2001). From the 1950s through the 1980s, the world's marine fisheries production increased an average of 6% per year. More recently, increases have been in the range of 0.6–1.5% (FAO 1999). The northwest Pacific remains the most important fishing area in terms of both volume and the value of its landings. For the world as a whole, landings of marine fish are leveling off.

Fisheries in the northwest, southeast, and east central Atlantic Ocean reached their maximum production levels one to two decades ago and are currently exhibiting a declining trend in total catches. In the northeast Atlantic, southwest Atlantic, west central Atlantic, east central Pacific, and northeast Pacific Oceans and the Mediterranean and Black Seas, annual catches seem to have stabilized, or are declining slightly, after having reached a maximum potential a few years ago (FAO 1999). Both the declining and flattening catch trends in these areas are consistent with the observation that these areas have the highest incidence of fully exploited fish stocks and of stocks that are overexploited, depleted, or recovering after having been depleted (FAO 1999).

The main regions where total catches still follow an increasing trend and where, in principle, some potential for increases still exists are the east and west Indian, west central Pacific, and northwest Pacific Oceans. These areas tend to have a lower incidence of fully exploited, overexploited, depleted, or recovering fish stocks, with relatively more under-exploited or moderately exploited stocks. However, these areas are also the ones with the largest incidence of stocks whose state of exploitation is unknown or uncertain and for which production estimates and stock assessments are less reliable (USDOC 1999).

Distant-water fisheries' production has declined sharply since 1990 mainly because of the loss of the state-sponsored fleets of the former Union of Soviet Socialist Republics (USSR). The country having the largest distant-water fleet is Japan, which in 1996 had total catches of 668,000 mt of fish by its distant-water fleet. Japan's presence on the ocean has declined since the 1970s, when Japan produced 2 million mt. Overall, the state of exploitation of the primary fish stocks has remained unchanged since the early 1990s (FAO 1999). The Food and Agriculture Organization (FAO) of the United Nations indicates that among these major fish stocks, 47% are fully exploited and are, therefore, producing harvests that have reached or are very close to their maximum limit, with little or no room for further expansion. About 18% of these major stocks are overfished with no room for expansion. In most of the latter cases, some remedial action is being taken to rebuild the fishery. Another 9% appear to be depleted with a loss of total production that affects society and local economies. One percent of the main stocks seem to be recovering (FAO 1999). The top seven wild-capture species that accounted for 25% of total production are (1) anchoveta, (2) Alaska pollock, (3) Atlantic herring, (4) skipjack tuna, (5) chub mackerel, (6) Japanese anchovy, and (7) Chilean jack mackerel (FAO 2000b).

WORLD AQUACULTURE STATUS

Aquaculture can be the answer for some of the world's future protein needs. On a worldwide basis, aquaculture has tremendous potential to supplement and/or replace wild captured species for human consumption and for export purposes. FAO defines aquaculture as the "farming of aquatic organisms, including fish, molluscs, crustaceans, and aquatic plants" (FAO 1995). Farming implies human intervention in the rearing process to enhance production practices such as stocking, feeding, health maintenance, and predator protection (FAO 2000b). Farming also implies individual or corporate ownership of the stock under cultivation.

The latest data available from FAO indicate that in 1999 global aquaculture production reached 33.31 million mt (Table 1.1). Since 1984, aquaculture has enjoyed an annual growth rate of 11%. FAO estimates

TABLE 1.1. Aquaculture Production of Animal Products, 1984–1998, by Economic Class

Year	Production (Kmt)			
	World	Industrialized Countries	Nonindustrialized Countries	Nonindustrialized as % of World Total
1984	6.89	2.19	4.70	68.3
1985	8.03	2.33	5.70	71.0
1986	9.16	2.48	6.68	72.9
1987	10.57	2.61	7.97	75.4
1988	11.66	2.70	8.97	76.9
1989	12.28	2.75	9.53	77.6
1990	13.04	2.85	10.19	78.1
1991	13.72	2.77	10.95	79.8
1992	15.45	2.76	12.70	82.2
1993	17.86	2.78	15.08	84.4
1994	20.75	2.83	17.93	86.4
1995	24.50	3.02	21.48	87.7
1996	26.76	3.10	23.66	88.4
1997	28.74	3.21	25.53	88.8
1998	30.74	3.40	27.35	88.9

Source: FAO 2000a.

TABLE 1.2. Aquaculture Production of Animal Products in 1998, by Region

Region	Production (Kmt)	% of Total
Asia	27.34	88.6
Europe	1.95	6.3
North America	0.66	2.1
South America	0.60	1.9
Africa	0.19	0.6
Oceania	0.12	0.4
Total	30.86	

Source: FAO 2000a.

that one of three fish consumed is now farm raised. Nearly 90% of all aquaculture is produced in Asia according to the FAO database (Table 1.2). Economically, the value of global aquaculture is some US$52 billion (FAO 2000b). With the exception of mainland China, the average food fish supply for the world is close to the levels of the early 1990s, which is somewhat lower than that of the 1980s. To feed future world populations and allow for localized production of fish and shellfish for individual consumption, aquaculture must continue to grow between 10 and 15% per year. This growth, however, will be dependent on local country resources, their need for additional protein, and regulations that are nonrestrictive and encourage development.

The production of finfish is the dominant global aquaculture activity and accounts for 49% of total production and 55% by value. Chinese and Indian carp account for the greatest share (42%) of this total. A key factor in this rapid growth is the increasing availability of hatchery-produced seed. Finfish account for 99% of freshwater aquaculture production, but less than 10% of the species are cultured in the marine environment (Table 1.3).

The top 10 cultured aquatic species on a worldwide tonnage basis are:

1. Kelp
2. Pacific cupped oyster
3. Silver carp
4. Grass carp
5. Common carp
6. Bighead carp
7. Yesso scallop
8. Japanese carpet shell
9. Crucian carp
10. Nile tilapia

Approximately 81% of the world's total finfish, shellfish, and aquatic plant production originates in low-income, food-deficit countries (LIFDC) (FAO 1999). Most of this production comes from six countries (China, India, the Philippines, Indonesia, Taiwan, and Bangladesh) with China accounting for approximately 85% of the total (FAO 1999). Figure 1.1 shows the major aquaculture producing regions in the world. It is interesting to note that the majority of current aquaculture production occurs north of the equator.

Tables 1.4 through 1.10 provide specific country and regional information on aquaculture production volumes and values.

PUBLIC, ANIMAL, AND ENVIRONMENTAL HEALTH ISSUES

In extensive aquaculture production systems, the cultured organisms rely solely on available natural foods (e.g., plankton, detritis, seston, etc.). Semi-intensive aquaculture involves fertilization to enhance the level of natural foods in the aquaculture system and/or the use of supplemental feeds (e.g., low-protein feeds from locally available plants or

TABLE 1.3. Production (mt) of Aquacultured Animal Products in 1998 by Species Groups and Economic Status of Country

Species	World			Nonindustrialized Countries			Industrialized Countries		
	Tonnage	Subtotal	% of Total	Tonnage	Subtotal	% of Total	Tonnage	Subtotal	% of Total
Freshwater Fish									
Carps, barbels and other cyprinids	14,142,298			13,926,847			215,451		
Miscellaneous freshwater fish	2,240,182			956,599			16,174		
Tilapias and other cichlids	972,773			190,586			31,961		
River eels	222,547			12			2,022		
Sturgeons, paddlefish	2,034	17,579,834	57.0	1,971,658	17,045,702	62.1	268,524	534,132	15.7
Diadromous Fish									
Salmons, trouts, smelts	1,291,444			318,429			973,015		
Miscellaneous diadromous fishes	393,021	1,684,465	5.5	388,136	706,565	2.6	4,885	977,900	28.7
Marine Fish									
Jacks, mullets, sauries	204,498			46,504			157,994		
Redfishes, basses, congers	201,380			43,506			157,874		
Flounders, halibuts, soles	33,445			22,703			10,742		
Tunas, bonitos, billfishes	5,140			0			5,140		
Cods, hakes, haddocks	148			0			148		
Miscellaneous marine fishes	336,455	781,006	2.5	327,268	439,981	1.6	9,187	341,085	10.0

Crustacea									
Shrimps, prawns	1,113,887			1,108,167			5,720		
Freshwater crustaceans	319,942			302,253			17,689		
Sea spiders, crabs	77,411			77,155			256		
Lobsters, spiny-rock lobsters	71			71			0		
Miscellaneous marine crustaceans	52,736	1,564,047	5.1	52,736	1,540,382	5.6	0	23,665	0.7
Mollusks									
Oysters	3,537,830			3,064,726			473,104		
Clams, cockles, arkshells	2,226,025			2,139,434			86,591		
Mussels	1,377,830			642,644			735,186		
Scallops, pectens	874,225			647,231			226,994		
Freshwater mollusks	13,131			13,131			0		
Abalones, winkles, conch	3,749			2,326			1,423		
Squids, cuttlefish, octopuses	33		29.6	0			33		
Miscellaneous marine mollusks	1,110,041	9,142,864		1,110,041	7,619,533	27.8	0	1,523,331	44.7
Other Products		110,791	0.4	99,519	99,519	0.4		11,272	0.3
Total Production		30,863,067			27,451,682			3,411,385	
% of World Production					88.9			11.1	

Source: FAO 2000a.

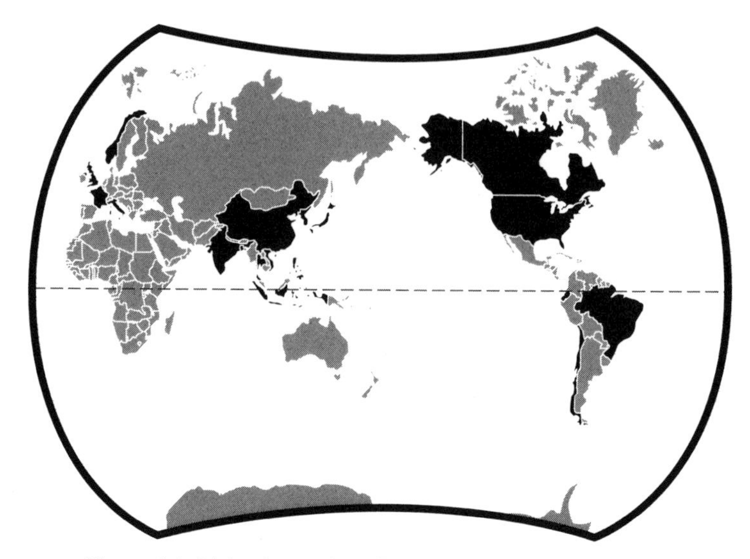

Figure 1.1. *Major Aquaculture Producing Regions in the World*

TABLE 1.4. Top Aquaculture-Producing Countries by Tonnage (mt)

Country—Rank	1996		1997		1998	
	Tonnage—K	US$—KK	Tonnage—K	US$—KK	Tonnage—K	US$—KK
China	22,208	21,311	24,030	23,549	27,072	25,449
India	1,783	2,057	1,862	2,142	2,030	2,223
Japan	1,349	5,019	1,340	4,706	1,290	4,126
Philippines	981	1,268	958	946	954	639
Indonesia	881	2,180	778	2,063	814	2,150
Korea, Rep. of	897	974	1,040	1,204	797	766
Bangladesh	450	1,223	513	1,404	584	1,494
Thailand	557	1,908	552	1,899	570	1,807
Vietnam	441	962	509	1,276	538	1,357
Korea, Dem. People's Rep.	782	441	489	308	481	303
United States of America	393	736	438	771	445	781
Norway	322	995	367	1,052	409	1,134
Chile	323	829	375	960	361	1,001
Spain	233	249	239	252	313	282
France	286	603	287	633	274	614
Taiwan	272	1,182	270	950	255	818
Italy	214	407	217	407	250	471
Ecuador	109	650	135	681	147	731
United Kingdom	110	269	130	427	137	428

Source: FAO 2000a.

TABLE 1.5. Top Aquaculture-Producing Countries by Tonnage (mt), Middle East, India Subcontinent, Far East, Central and SE Asia Regions

Country—Rank	1996		1997		1998	
	Tonnage—K	US$—KK	Tonnage—K	US$—KK	Tonnage—K	US$—KK
China	22,208	21,311	24,030	23,549	27,072	25,449
India	1,783	2,057	1,862	2,142	2,030	2,223
Japan	1,349	5,019	1,340	4,706	1,290	4,126
Indonesia	881	2,180	778	2,063	954	2,150
Philippines	981	1,268	958	946	954	639
Korea, Rep. of	897	974	1,040	1,204	797	766
Bangladesh	450	1,223	513	1,404	584	1,494
Thailand	557	1,908	552	1,899	570	1,807
Vietnam	441	962	509	1,276	538	1,357
Korea, Dem. People's Rep.	782	441	489	308	481	303
Taiwan	272	1,182	270	950	255	818
Myanmar	68	704	87	865	85	1,318
Iran, Islamic Rep. of	30	—	30	—	33	—
Pakistan	18	13	22	14	24	14
Israel	18	52	18	63	19	64
Cambodia	10	23	12	26	14	29
Laos, People's Dem. Rep.	14	36	14	35	14	35
Nepal	11	11	12	13	13	14
Syrian Arab Rep.	6	29	6	26	7	33

Source: FAO 2000a.

agricultural by-products). Intensive aquaculture systems use external high-protein prepared feeds (e.g., fish meal and fish oil based) as a food source for the cultured organisms.

Current aquaculture trends in nonindustrialized countries are principally twofold: (1) intensive culture of high-value species for export and revenue generation and (2) extensive culture of fast-growing species such as the carps, tilapia, and catfish to feed local populations. Nevertheless, as in industrialized countries, environmental and public health issues in nonindustrialized countries are usually addressed in a reactive and not a proactive manner.

Current aquaculture trends in industrialized countries include continued development of more intensive aquaculture systems for breeding, rearing, and harvesting of high-value species within carefully controlled grow-out facilities. More intensive, closed aquaculture pro-

TABLE 1.6. Top Aquaculture-Producing Countries by Tonnage (mt), Europe

Country—Rank	1996		1997		1998	
	Tonnage—K	US$—KK	Tonnage—K	US$—KK	Tonnage—K	US$—KK
Norway	321	995	367	1052	409	1134
Spain	232	249	239	252	314	282
France	286	603	287	633	274	614
Italy	214	407	217	407	250	471
United Kingdom	110	269	130	427	137	428
Netherlands	100	84	98	82	120	88
Germany	75	125	59	101	67	80
Russian Federation	53	142	60	162	66	174
Greece	40	236	49	246	60	275
Denmark	41	139	39	134	42	152
Ireland	35	83	37	77	40	81
Poland	28	69	29	60	30	62
Ukraine	33	83	30	78	28	76
Faeroe Islands	18	66	22	73	20	67
Czech Republic	18	49	18	47	17	46
Finland	18	58	16	48	16	50
Hungary	8	14	9	18	10	20
Romania	14	35	11	17	10	16
Portugal	5	31	7	48	8	51
Croatia	3	9	4	11	6	23

Source: FAO 2000a.

duction systems will be necessary to ensure a high-quality product, as well as to prevent pollutants from entering into the environment from aquaculture facilities. Such controls will require more attention to the subject of aquaculture biosecurity at the production level.

As with any developing and growing industry, unintended consequences will develop before the technology is fully matured. Over time, education, training, and risk-based planning will reduce the possibility of a negative event. For aquaculture to reach its full potential, it must have a central planning approach (i.e., an authority to develop a national sustainable aquaculture plan that is both environmentally safe and socially acceptable). Without a national policy, aquaculture primarily in industrialized countries (with few exceptions) will continue to develop slowly. These exceptions (Japan, Norway, Chile, and the USA farm-raised catfish industry) should be studied carefully to learn from some of their past unintended events.

TABLE 1.7. Top Aquaculture-Producing Countries by Tonnage (mt), Africa

Country—Rank	1996		1997		1998	
	Tonnage—K	US$—KK	Tonnage—K	US$—KK	Tonnage—K	US$—KK
Egypt	76	145	73	166	139	327
Nigeria	18	44	18	42	20	47
Madagascar	5	18	5	19	7	36
South Africa	3	8	4	9	5	13
Zambia	5	20	5	14	4	7
Tanzania	3	1	3	0.9	3	0.9
Morocco	2	12	2	9	2	8
Tunisia	1	7	2	10	2	9
Sudan	1	1	1	2	1	2
Congo, Dem. Rep. of	0.8	0.8	0.8	1	0.8	0.8
Seychelles	0.3	3	0.6	7	0.6	7
Ivory Coast	1	2	0.4	0.8	0.4	0.8
Ghana	0.5	1	0.4	0.7	0.4	0.7
Uganda	0.2	0.2	0.4	0.3	0.4	0.3
Algeria	0.3	0.9	0.3	0.9	0.3	0.8
Malawi	0.2	0.2	0.2	0.2	0.2	0.2
Zimbabwe	0.2	0.7	0.2	0.7	0.2	0.7
Kenya	0.6	1	0.2	0.3	0.2	0.2
Rwanda	0.1	0.2	0.1	0.2	0.1	0.2
Congo, Rep. of	0.1	0.3	0.1	0.3	0.1	0.3
Libyan Arab Jamahiriya	0.1	0.1	0.1	0.1	0.1	0.1

Source: FAO 2000a.

AQUACULTURE ISSUES

The aquaculture sector is not without problems. Listed below are some of the identified issues associated with global aquaculture. These issues will be further addressed and discussed in Chapters 2 to 6 of this book.

1. The positive societal consequences of aquaculture must be publicly understood so that misleading negative perceptions do not discourage development. Job creation, fulfillment of food requirements, possible stock enhancement, and the effect on a local or national economy should be publicly acknowledged before detractors react to aquaculture. Aquaculture must not be viewed as a nuisance (Baily 1988; Chua et al. 1987).
2. Industrialized countries go to great lengths to test drugs and their effectiveness, whereas some nonindustrialized countries do not. Antibiotic resistance to certain drugs is a current global

TABLE 1.8. Top Aquaculture-Producing Countries by Tonnage (mt), South America and Caribbean Region

Country—Rank	1996		1997		1998	
	Tonnage—K	US$—KK	Tonnage—K	US$—KK	Tonnage—K	US$—KK
Chile	323	829	375	960	361	1002
Ecuador	109	650	135	681	146	731
Brazil	78	305	88	328	95	348
Colombia	30	138	44	190	46	191
Cuba	34	30	46	41	38	34
Venezuela	7	24	9	28	11	33
Panama	5	32	7	36	10	53
Costa Rica	7	36	7	36	9	38
Honduras	10	69	9	64	8	57
Peru	7	54	8	48	8	44
Nicaragua	3	17	4	21	5	25
Guatemala	3	15	4	21	3	15
Belize	1	6	1	10	2	13
Argentina	1	9	1	9	1	9
Dominican Republic	0.8	3	0.7	2	0.8	2
Bolivia	0.4	1	0.4	1	0.4	1
El Salvador	0.4	0.9	0.4	1	0.4	3
Puerto Rico	0.1	0.4	0.1	0.4	0.1	0.4
Paraguay	0.4	0.6	0.4	0.6	0.1	0.2

Source: FAO 2000a.

TABLE 1.9. Top Aquaculture-Producing Countries by Tonnage (mt), North American Region

Country—Rank	1996		1997		1998	
	Tonnage—K	US$—KK	Tonnage—K	US$—KK	Tonnage—K	US$—KK
United States of America	393	736	438	771	445	781
Canada	71	283	82	322	91	341
Mexico	31	73	39	121	41	156

Source: FAO 2000a.

regulatory problem. With limited sales to the aquaculture sector, drug companies are reluctant to undertake the necessary research and regulatory submissions for fish health approvals.

3. Detractors call aquaculture "biological pollution." The accidental or deliberate introduction into marine ecosystems of exotic or genetically manipulated species of micro- or macroorganisms must be considered in terms of risk management strategies.
4. Because raising animals in confined spaces contributes to the susceptibility to disease, the subject of biosecurity of aquaculture

TABLE 1.10. Top Aquaculture-Producing Countries by Tonnage (mt), Oceania

Country—Rank	1996		1997		1998	
	Tonnage—K	US$—KK	Tonnage—K	US$—KK	Tonnage—K	US$—KK
New Zealand	75	46	77	46	94	57
Australia	25	169	27	163	28	164
Fiji Islands	0.2	1	0.3	1	10	2
Kiribati	5	2	5	2	7	0.6
New Caledonia	1	11	1	12	2	18
Guam	0.2	1	0.2	0.1	0.2	0.8
French Polynesia	0.07	0.5	0.06	0.4	0.05	0.4
Papua New Guinea	0.02	0.07	0.03	0.08	0.03	0.01
Solomon Islands	0.01	0.08	0.01	0.01	0.01	0.02
Cook Islands	0.005	0.005	0.005	0.005	0.005	0.005
Micronesia	0.005	0.005	0.005	0.005	0.005	0.005
Samoa	0.005	0.005	0.005	0.005	0.005	0.005

Source: FAO 2000a.

sites has become an important consideration in maintaining the health and growth of farm-raised species.

5. The environmental effects of organic waste from fish farming must be considered. Although the pollution load may be substantial, the pollution effects are restricted to the immediate vicinity of the fish farm. On-site treatment is possible, or in high tidal areas, the tidal flush may be sufficient to disperse the potential problem (Stechey 1988). Responsible stewardship requires that proper site selection is the first consideration when planning an aquaculture operation.
6. Environmentally, the phosphorus content of aquaculture feeds must be reduced or means must be developed to reduce their release into the surrounding environment.
7. Aquaculture operations, regardless of their location, must have an adequate supply of good-quality water. Proper investigation of potential aquaculture sites before investment must be given first consideration in the development of a business plan.
8. A framework must be established for the leasing of public resources in coastal areas. Concerns over public access, use conflicts, and suitable sites must be resolved before aquaculture programs can be used by the private sector (DeVoe and Mount 1989).

9. Issues in mussel and cage culture such as mooring and cage strength (when located in highly turbulent coastal waters), control of predators, and large-scale seed and feed requirements must be taken into account (Appukuttan 1988).
10. High-value transgenic species will be developed that will be capable of using vegetable-based protein diets thus reducing the risk that we are fishing the oceans to feed farm-raised species.

THE FUTURE OF AQUACULTURE

The phrase "catch the wave" is an appropriate expression for the future of aquaculture. Aquaculture is on the wave but not as yet on its crest. Recent global headlines listed below illustrate the future potential of sustainable aquaculture.

Fish farms may become No. 1 food source—fastest growing sector of world economy
World Watch Institute, Washington, DC, Oct. 3, 2000

Fish farming, the fastest growing sector of the world food economy, will surpass cattle ranching as a food source by the end of this decade, the Worldwatch Institute predicted. Aquaculture output, growing by 11% a year, has climbed from 13 million mt of fish produced in 1990 to 31 million mt in 1998, according to an analysis written by Lester Brown, chairman of the Institute, which tracks environmental trends around the world. While meat production is concentrated in industrialized countries, about 85% of fish farming is in developing nations. The world's leader is China, where aquaculture began 3000 years ago. China accounts for more than two-thirds of world aquaculture output.

Pennsylvania hooked on fish farming industry—steady markets to lead to growth in industry
The *Patriot-News*, Harrisburg, PA, Oct. 4, 2000

Fish farming—whether raising fish to stock ponds, feed the hungry or fill aquariums—is big business in a state better known for its horses, cows, and hogs. "We have about 150 fish farms in 64 of Pennsylvania's 67 counties," Leo Dunn, aquaculture coordinator for the State Department of Agriculture, said during the Pennsylvania Aquaculture Conference at the Radisson Penn Harris Hotel and Convention Center. Pennsylvania ranks 17th nationally in aquaculture," Dunn said. "Lancaster County has the largest number of fish farms of these counties—nine."

China No. 1 global seafood producer
China Online, Beijing, China, Nov. 22, 2000

According to sources from the Third World Fishery Conference, China has ranked first in the world in seafood output for 10 consecutive years since 1990, as its fishery industry has markedly enhanced its overall production capacity. China's seafood production reached 41.22 million mt in 1999, an increase of more than seven times over the 1978 figure. Of that total output, 23.96 million mt came from aquaculture, or water plants and animals that are regulated and cultivated for human use or consumption, the Nov. 16 *Hua Sheng Bao* (*Hua Sheng Overseas Chinese Newspaper*) reported. This makes China the only country in the world to reap more seafood from aquaculture than from capture fisheries, the article said. China's per capita share of seafood production went up from 4.8 kilograms (10.56 pounds) in 1978 to 32.7 kilograms (72.1 pounds) in 1999, which is above the world average.

El Rosario net $8 million shrimp loan—company seeks funds for aquaculture expansion, environmental upgrades
Business News Americas, Guayaquil, Ecuador, Dec. 20, 2000

The Inter-American Investment Corporation (IIC) has approved a seven-year US$6 million "A" loan and a nine-year income participating US$2 million "C" loan for Ecuador Guyayaquil-based shrimp company El Rosario Camarones Humboldt Panama. The loan will help with environmental upgrades and biosecurity for the company's US$37 million aquaculture project that entails expanding the El Rosario shrimp farm operation to 4196 hectares (ha) from 3796 ha of grow-out and nursery ponds in response to increasing shrimp demand.

Florida firm to build new "closed aquaculture" facility—new facility to grow tilapia in Malaysia
Business Wire, Lakeland, Florida, Dec. 20, 2000

Sheffield Aquaculture, Inc., a wholly owned subsidiary of Sheffield Environmental Services, Inc., of Lakeland, Florida is developing the first of its kind "Closed Aquaculture" facility on 150 acres of land located in Sabah, Malaysia. Thermotek International, Inc. (TTKI) of Burlington, Iowa is being contracted to install the water treatment portion of the organic facility. This system is estimated to cost $2.5 million dollars. This "Closed Aquaculture" project will produce over 30,000 pounds of organically raised Tilapia (very fast growing tropical fish) weekly. Long-term contracts from total production are being negotiated in the Pacific Rim. Four additional "Closed Aquaculture" facilities are being planned worldwide. TTKI will provide all water treatment installations for these projects generating in excess of $100 million in revenue.

U.S. biologists to use Red Sea to turn desert into shrimp farm
Associated Press, Asmara, Eritea, Dec. 6, 2000

The Americans in shorts and baseball caps dip their hands in chlorine before plunging them into the waist-high circular concrete tanks filled with seawater on the deserted stretch of the Red Sea beach. The shrimp inside the tanks are their prime investment in a sea farming project that stands to make a lot of money for this tiny, impoverished nation in the Horn of Africa. "We're reversing the flow of water, which runs from the mountains and washes the soil into the sea, by recapturing nutrients and putting them back into the land," said Ned Daugherty, a university researcher and the environmental architect at Seaphire International's first African sea farm. In 1999, Daugherty and his associates set up Seawater Farms Entree, a joint venture between the Phoenix, Arizona-based Seaphire and the Eritrean government. They began by laying miles of pipe across the beach to pump seawater onto 1482 acres of desert. The pipe that runs past a burned-out T-55 tank delivers seawater to 230 brick and concrete tanks lined up in neat rows. Each tank is home to 200,000 post-larvae shrimp that were flown to Eritrea from a Seaphire hatchery in Mexico. The Eritrean farm is building its own hatchery stocked with selected shrimp that take six to eight months to mature. After passing through the shrimp tanks, the seawater then flows into the farm's three artificial salt lakes, where tilapia, a tropical fish, are bred for export. Tilapia fish meat is also used to make shrimp feed, and the fish's skin can be turned into attractive leather.

Chilean farmed salmon exports up 29%, Europe increased purchases by 66%, U.S. markets buy 43% more
World Catch News Network, Santiago, Chile, Jan. 5, 2001

Further evidence of the explosive growth in Chilean farmed salmon exports was announced by the Association of Salmon and Trout Producers, which reported a 29% increase in shipments abroad for the first 11 months of 2000, compared with the same period in 1999. Sales of farmed salmon exports were recorded at US$823 million, about US$195 million more than during the same period in 1999, according to *Finning News*, and close to US$57 million more than the overall export figure for 1999. The association suggested that once year-end statistics are added, its target of US$950 million in exports for 2000 will have been achieved, the report said. Export volumes for the first 11 months of the year increased by 39% to 180,000 mt, compared with the same period in 1999, when about 155,000 mt were shipped abroad. The strongest markets for Chilean exports are Europe, which, according to the report, increased its purchases by 68%; Latin America, which increased purchases by 38%; and the United States, with an increase of 43%.

Technology drives vision of deep-sea farming
Canadian Press, Bedford Nova Seotia, Canada, Jan. 15, 2001

A new era is emerging for some Atlantic fishermen, one in which the traditional hunt for seafood is shifting to mapped, undersea farms where trawls always make a catch. Eric Roe, an executive with Clearwater Fine Foods, a fishing giant on the East Coast, says new technologies are heralding nothing less than a "revolution in fisheries management." Combined with software developed by Canadian marine geologists, the company can use the US$3 million worth of data to figure out exactly where the sea life is likely to flourish on the Georges' Bank off Nova Scotia, one of the richest fishing grounds in the country. As ocean bottoms are mapped, the possibility of deepwater aquaculture is becoming a reality. That shift means the US$100 million scallop fishing industry could evolve from a hunt based on sea captains' memories to something more like ranching.

Sea Land diversifies into fish farms
The *Nelson Mail*, Nelson, Australia, Jan. 23, 2001

Sealord Group is poised to get involved in fish farming, a new field for the Nelson-based seafood company. The venture is one of a number to be developed as a result of a change in the company's ownership, approved by the government yesterday. Under the new ownership, Japanese seafood giant Nissul and the Treaty of Waitangi Fisheries Commission will jointly buy Brierly Investments' 50% shareholding in Sealord for just under US$208 million.

Cherrystone to expand processing facility for aquacultured clams
World Catch News Network, Cheriton, Virginia, Feb. 5, 2001

Cherrystone-Aqua-Farms of Cheriton, VA, announced that construction has begun on a 6000 square foot addition to its processing facility. The expansion will accommodate additional packing lines, coolers, packaging storage, and offices. "The new facility will allow Cherrystone to reach its goal of doubling its current sales to 100 million clams per year, while operating in a facility designed for its highest standards of sanitation," said Tim Parsons, director of sales and marketing. "It will also allow Cherrystone to expand both its offerings of pack sizes and private branding programs."

Midwestern farmers to try fish farms to make up for crop losses
St. Louis *Post Dispatch*, St Louis, Missouri, Feb. 7, 2001

Three miles southwest of Dietrich's lake, Duane and Allen Waeltz spent last spring and summer—when not tending to their row crops—building an 8800 gallon recirculating water system inside a former hog barn.

While the temperature outside is often below freezing, Waeltz is burning propane at triple last year's price to keep the operation, known as raceway, at a constant 80 degrees. The depressed economies of coal and crops have spurred farmers and others to consider their options. The prospect of a better return on investment has brought aquaculture to southern Illinois.

Farmed salmon firm keeps growing—Salmones Multiexport purchase of fish farm to boost production
World Catch News Network, Osorno, Chile, Feb. 16, 2001

Chilean farmed salmon producer, Salmones Multiexport, has purchased 98.5% of all shares in fish farm operator, Rio Bueno, owned by the Momberg Bohorques family, for approximately US$6 million, according to a report by *Finning News*. The purchase will reportedly enable Multiexport to increase its salmon and trout production from 25,000mt to more than 80,000mt annually. Multiexport recorded export sales of US$65.7 million last year, and was the second largest exporter of farmed salmon behind Marine Harvest, which exported US$78 million worth of farmed salmon in 2000.

Africa's $12 million shrimp farm
Africa News Service, Beira, Mosambique, Feb. 21, 2001

Approximately US$12 million has been invested in a prawn farming project in the central Mozambican port city of Beira. The Mozambican news agency, AIM, quoted the country's Fisheries Minister, Cadmiel Muthemba, as saying the project, a joint venture between Mozambican and Chinese businessmen, was formally launched last week. He added that "Sol-Mar" expects to farm an average of 800mt of prawns a year, mainly for export. A similar project will start this year in Quellmane, the capital of the central province of Zambezia, he said, adding that preparations for the launching of the project, including the setting up of equipment and other infrastructure necessary for the farming of prawns, were underway.

CONCLUSION

As aquaculture assumes a greater role in the supply of food, techniques such as HACCP principles, best management practices, and risk analysis, comprised of risk assessment, management, and communication, are necessary to minimize the unintended consequences from this technology. Aquaculture endeavors should involve an integrated approach that builds on the totality of issues and past experiences so that aquaculture neither causes nor gives the appearance of contributing to unacceptable risks to public health or other harm that negates the improved economic and nutritional benefits that the technology offers.

REFERENCES

Appukuttan, K.K. 1988. Present status and problems of mussel culture in India. J. Indian Fish Assoc. 18: 39–46.

Bailey, C. 1988. The social consequences of tropical shrimp mariculture development. Ocean Shoreline Mgt. 11(1): 31–44.

Chua, T.-E., L.M. Chou, and A.M.B.H. Jaafar. 1987. Coastal resources management issues and plan formulation. ICLARM Tech Report. No.18, pp. 154–167.

DeVoe, M.R. and A.S. Mount. 1989. An analysis of ten aquaculture leasing systems: Issues and strategies. J. Shellfish Res. 8(1): 233–239.

Ess, C. 2001. 82 Fish species at risk of extinction. Natl. Fisherman 81(10): 12–14.

FAO. 2000a. Fishstat+, version 2.3. Downloadable database from www.fao.org/fi/statist/fisoft/fishplus.asp. Rome, Italy: Food and Agriculture Organization of the United Nations.

FAO. 2000b. The state of the world fisheries and aquaculture 2000. Food and Agriculture Organization of the United Nations. ISBN 92-5-104492-9. Rome, Italy.

FAO. 1999. The state of the world fisheries and aquaculture 1998. Food and Agriculture Organization of the United Nations. ISBN 92-5-104187-3. Rome, Italy.

FAO. 1995. Code of conduct for responsible fisheries. Food and Agriculture Organization of the United Nations. Fisheries Department. p. 41. Rome, Italy.

Stechey, D. 1988. Factors influencing the design of effluent quality control facilities for commercial aquaculture. Proceedings of Aquaculture International Congress, p. 54. Windsor, Ontario, Canada.

USDOC. 1999. Special Summer of 1999 Aquaculture Workshop Report. NOAA, Silver Spring, MD.

2

Aquaculture Associated Public, Animal, and Environmental Health Issues in Nonindustrialized Countries

Peter Howgate, Stuart Bunting, Malcolm Beveridge, and Alan Reilly

INTRODUCTION

FAO fishery statistics (FAO 2000) show that in 1998 aquaculture production consisted of 30.9 million mt of animal products, which provided 32% of the world's supply of fish for human consumption (Chapter 1, Table 1.2). Since 1984 (the earliest year for which detailed FAO statistics for aquaculture are available) production has increased approximately 11% per year, thus making aquaculture one of the

Public, Animal, and Environmental Aquaculture Health Issues,
Edited by Michael L. Jahncke, E. Spencer Garrett, Alan Reilly,
Roy E. Martin, and Emille Cole.
ISBN 0-471-38772-X (cloth)

fastest-growing food production systems in the world. On the other hand, supplies of fish for human consumption from capture fisheries have shown almost no increase since the end of the 1980s. In the decade of 1988 to 1998, the world's population increased by about 15%. However, total world fish production exceeded this figure; thus the per capita consumption of fish increased from about 13.6 kg/person/year, live weight basis, to about 16.0 kg/person/year. These increases in total fish production and in per capita consumption were made possible by the increased production of fish from aquaculture rather than from capture fisheries. Should aquaculture production continue to grow at around 11% per annum, it will overtake capture fisheries in the production of fish for human consumption by the year 2006. It would be unwise to assume the 11% growth rate can be maintained in the future, but it is clear that aquaculture will be crucial for maintaining, if not increasing, supplies of fish for an increasing world population. Additionally, commercial aquaculture commodities contribute significantly to the economies of many nonindustrialized countries, where high-value species such as shrimp are a major source of foreign exchange.

This chapter considers public, animal, and environmental health issues associated with aquaculture production in nonindustrialized countries. There is increasing concern in nonindustrialized countries regarding possible risks to the environment, animals, and public health from aquaculture. Examples of such concerns are the destruction of mangrove areas for shrimp ponds, the use of antibiotics to control diseases in cultured animals, the use of raw human and animal manures as pond fertilizers, and the spread of viral and bacterial pathogens from cultured fish to wild stock.

AQUACULTURE PRODUCTION

Most of the world's aquaculture production is, and has been, located in Asia, according to the classification of FAO's database (Chapter 1, Table 1.2). This region has increased its relative contribution to world supplies of animal products from aquaculture from 76% in 1984 to 89% in 1998. Countries classified as nonindustrialized (or developing) in the FAO classification dominate world aquaculture production by contributing 90% of the total aquaculture products for human consumption in 1998. This proportion has increased steadily from 68% in 1984. Chapter 1, Table 1.3 shows the breakdown of the total world aquaculture production by biological and environmental groupings and is separated into industrialized and nonindustrialized countries. On a world basis, vertebrate fish raised in the freshwater environment dominate aquaculture production. Finfish is not nearly as important for industrialized countries as it is for nonindustrialized countries. Shrimp and other crustaceans, although only 5.6% of the total production in nonindustrialized countries, are economically important because much of the production is exported. In addition to these animal products, 8.57 million mt of red and green seaweeds were produced by aquaculture in 1998, mostly in Asia.

Given the present, and likely, increasing importance of aquaculture products in supplying fish for human consumption, it is necessary to consider whether or not there are any aspects of aquaculture production that could introduce new public health hazards or alter the risks of known hazards of fishery products. The following review and discussion will indicate that the public health hazards associated with aquaculture products are influenced by a wide range of factors, including species and production systems, the environment in which culture is undertaken, and the geographic region of production.

OVERVIEW OF AQUACULTURE PRODUCTION SYSTEMS AND METHODS

Introduction

As previously stated, nonindustrialized countries account for the majority of global aquaculture production, most of which occurs in Asia, with Africa and Latin America accounting for less than 2% and 1%, respectively (see Chapter 1, Table 1.2). The reasons for such geographic disparities are complex and inextricably linked with prevailing historic cultural, political, and economic conditions.

The main aquaculture production systems utilized by the industry are summarized in Chapter 1, Table 1.3. Following the terminology of Coche (1982) and many other authors, aquaculture production methods are classified as extensive, semi-intensive, or intensive. In extensive aquaculture, the animals rely solely on available natural food, such as plankton, detritus, and seston. Semi-intensive aquaculture involves fertilization to enhance the level of natural food in the systems and/or the use of supplementary feed. Such feeds are usually low protein (generally less than 20%), are a mixture of locally available plants or agricultural by-products, and complement the intake of natural food. In intensive aquaculture, animals are almost exclusively reliant on an external supply of high-protein food (usually greater than

20%), generally based on fish meal and fish oil. All mollusk and seaweed culture and approximately 80% of global production of farmed finfish and crustaceans take place within extensive and semi-intensive farming systems. This section briefly summarizes the prin-cipal species, systems, and methods in a global context. Given its overwhelming importance, the focus is very much on Asia and although many tropical fish, invertebrate, and plant species are farmed, the focus is on the salient features of the most important species groups.

Coastal Aquaculture

Fish Some 1.2 million mt of marine and brackish water fish are produced in farms in Asia. Production consists largely of two species, milkfish (*Chanos chanos*; 400,000 mt) and yellowtail (*Seriola quinqueradiata*; 140,000 mt), with limited but increasing quantities of sea bream (*Pagrus major*), barramundi or giant sea perch (*Lates calcarifer*), and groupers (*Epinephelus* spp.) (Beveridge and Haylor 1998). Production of herbivorous/planktivorous milkfish and mullet (Family: Mugilidae) reared by traditional methods in coastal fishponds in China, Indonesia, and the Philippines still dominates their industry. Over the past two decades, however, coastal fishponds have been converted to shrimp production, although ironically, in Ecuador and elsewhere in Latin America, disease has forced a number of shrimp producers to switch to rearing tilapia (*Oreochromis* spp.).

Intensive cage farming of carnivorous sea bass, sea bream and groupers in sheltered inshore coastal areas in Thailand, Malaysia, Singapore, Vietnam, and China is perhaps the most dynamic sector of marine finfish culture in nonindustrialized countries. The product from such efforts, in large measure, attempts to satisfy the live fish market. Production of these species now totals approximately 0.2 million mt. Cages are still largely built from locally available materials, although there is also a growing interest in utilizing European and North American industrial offshore designs that may allow for more exposed sites to be exploited. Production is primarily based on using trash fish as a feed source, although formulated feeds are beginning to be utilized.

Shrimp Shrimp postlarvae are derived either from wild fisheries, from wild-caught egg-bearing females or from hatchery-reared females. Postlarvae are then stocked in coastal ponds at a range of densities. They are grown for 3–4 months until they reach marketable size (18–35 g). Shrimp farming is frequently practiced in extensive or semi-intensive culture systems with low stocking densities (1–10 postlarvae m^{-2}), incorporating fertilizers and the use of low-cost, locally made feeds. Intensively managed shrimp farms have stocking densities ranging from 20 to $100 m^{-2}$ and are heavily reliant on pelleted, nutritionally complete feeds. These intensive systems use aeration and pumped seawater to replenish dissolved oxygen and flush away potentially harmful waste metabolites (Primavera 1998).

FAO statistics on global farmed shrimp production show tremendous increases during the 1980s with a much more modest growth in the 1990s. However, a more detailed analysis of the figures reveals that, although production in several countries such as China, Mexico, and Brazil began to accelerate during the 1990s, the decade was characterized by faltering production in other countries. Intensification, compounded by overcrowding of farms in many areas, poor husbandry, weak quarantine regulations, and poor enforcement of animal health regulations led to the promotion and spread of pathogens and outbreaks of disease that all but destroyed shrimp farming in many places. There are strong international initiatives through the FAO, the National Aquaculture Centres in Asia (NACA), and other organizations that encourage sound best management practices, usually based around the adoption of less intensive culture methods; nevertheless, disease still remains a problem for the industry.

Mollusks Mollusks, including those species reared for pearl cultivation, are generally farmed in intertidal and sheltered inshore coastal areas. Mussels are grown using spat collected from the wild, in bundles of bamboo or coir rope. In traditional or artisanal culture systems, there is little further management needed other than thinning of animals as they grow. However, despite the manpower, expense, and loss of spat, many farms re-lay spat, redistributing animals to more appropriate substrates and locations for increased growth. Off-bottom, grow-out methods, e.g., bamboo poles, racks, ropes and longlines, are generally preferred. In Thailand, the grow-out period for 5- to 7-cm green mussels, *Mytilus smaragdinus*, varies from 6 to 12 months with yields (meat and shell) ranging from 150 to $500 t ha^{-1}$. Yields in India have been reported at $1800 t ha^{-1}$ (Joseph 1998).

Oyster culture in nonindustrialized countries is also almost exclusively dependent on wild-caught spat. Because of problems of silting and heavy mortalities (with the exception of *Crass ostrea virginica* culture in the Bay of Mexico), off-bottom culture methods prevail. Hard cultch materials, such as oyster or mussel shell, concrete slabs, tiles, and sticks are used to collect spat in intertidal areas. The cultch (substrate) plus spat is then relayed to deeper areas of the intertidal zone, where the oysters will remain until they reach marketable size (7–9 cm; 6–7 months) (Joseph 1998).

Farming of pearl oysters, especially in the south Pacific, has been growing steadily. Production is based on Japanese methods, with pieces of shell and mantle inserted into the gonads of adult *Pinctada* species. Oysters with pearl implants are then transferred to lantern nets or cages, where they remain for an additional 2 years until the animals and pearls are harvested. Meat from pearl oysters does not generally enter the human food chain.

Seaweed Seaweed farming using fixed off-bottom lines, floating rafts, or longlines is practiced in open coastal areas and occasionally in ponds. Areas with extensive, level areas of sea bottom that are not exposed at low tides and with low freshwater inputs, reasonable nutrient levels, and high water transparency are ideal. Vegetative cuttings from the previous harvest or from a nearby farm generally serve to initiate culture of the carrageenan-bearing red seaweeds such as *Eucheuma denticulatum* and *Kappaphycus alvarezzi*, although spore-generated seed stock and seedlings are increasingly used in other seaweed cultures. With the exception of China, where some fertilization is used (Phillips 1998), plants rely on environmental nutrient concentrations to sustain growth.

Inland Aquaculture

The majority of global aquaculture production is located inland and remains dominated by pond-based fish culture, although practices such as rice-fish culture and cage-based culture are often important on a regional basis. Freshwater prawn farming is of local importance only, although in Thailand, toward the end of the last century, the controversial practice of growing penaeid shrimp in brackish water-filled inland ponds became popular (Flaherty et al. 2000).

Fish Rice-fish culture is still practiced in traditionally managed rice fields and is one of the oldest methods of freshwater fish farming. This type of farming often involves little more than management of the paddy and water to ensure that fish enter the fields from surrounding rivers or canals and that the environment is conducive to their growth. Unless stocking is practiced, fish are small and yields are low ($\sim 0.5\,t\,ha^{-1}\,y^{-1}$). Harvested fish tend to be eaten by the farmers, their families, and farm laborers.

Pond aquaculture accounts for the majority of farmed fish production from nonindustrialized countries. Traditional inland pond aquaculture practices are integrated with other agricultural practices and utilize local supplies of surface or groundwater. Production is relatively efficient, based on the use of on-farm or locally available crop and/or animal production by-products and household wastes. With stocking densities of one to five fish m^{-2}, often several carp or local fish species are stocked together, and typical yields range from 2 to 5 t ha y^{-1}. These ponds enhance habitat and landscape diversity and are integrated with the rural, social, economic, and natural environments (Beveridge and Phillips 1993; Edwards 1993; Beveridge et al. 1997). Indeed, in the most highly integrated systems, fishponds play a pivotal role in supporting other activities including water conservation (Pullin and Prein 1995).

Despite the apparent benefits associated with traditional pond aquaculture, two emerging trends are threatening the continued operation of these culture systems in rural areas. The first is the growth of aquaculture in periurban areas, capitalizing on waste effluents and ready access to markets. The second is increased production intensification caused by the gradual break with agriculture, the rise in land and water prices, especially in periurban areas, and the strong demand for fish by the ever-expanding urban masses. These trends have resulted in the gradual abandonment of traditional fish polycultures in favor of the monoculture of general species such as tilapia, eels, prawns, and terrapins destined for export markets (Wong Chor Yee 1999).

Inland cage fish farms are usually sited in multipurpose, publicly owned water bodies and are exploited for food and/or used to dissipate and/or assimilate waste materials. Cage farms are rarely integrated to the same degree as ponds. Many farms are initially heavily dependent on the supply of natural food, but there is an increasing reliance on supplemental foods, resulting in an increased exposure to financial risks (Beveridge 1984, 1996; Sevilleja et al. 1993; Santiago 1995; Beveridge and Stewart 1998).

In contrast with mariculture, seed for freshwater fish farming is largely hatchery derived. In nonindustrialized countries, carp species still account for the vast majority of freshwater fish production. Tilapia production is increasing rapidly and is expected to soon exceed 1 million mt per annum. In Lake Kariba, Africa, Costa Rica, and Jamaica, intensive pond and cage production of tilapia has been developing rapidly to satisfy new and expanding export markets to North America and Europe (Beveridge and McAndrew 2000).

Prawns According to the FAO, annual farmed production of the giant freshwater prawn *Macrobrachium rosenbergii*, principally in Asia, has shown a general increase from 10,000 mt in 1984 to 103,000 mt in 1998. Expansion of the industry is constrained by limited markets. Hatchery production is straightforward; mating takes place after female adults molt, and eggs are extruded within 24 h. Eggs are laid in freshwater and hatch within 24 h. Larvae must be transferred to water with a salinity of 12–16 ppt and kept for 3 to 7 weeks until they metamorphose into postlarvae. The postlarvae can then be moved to freshwater ponds for grow-out. *Macrobrachium* are generally grown under semi-intensive culture conditions (fertilization and feeding) with stocking densities ranging from 12 to 30 ha^{-1}. With a continuous cropping scheme, yields of 3–4 t ha y^{-1} can be produced (New and Singholka 1985).

FOOD SAFETY ISSUES ASSOCIATED WITH FISH AND FISHERY PRODUCTS

The hazards present in farmed fish and shellfish, with the exception of one class of hazard, are similar to those present in wild stocks of fish and shellfish. The exception is the presence of chemicals for the treatment of diseases in the farmed products. The human health hazards from veterinary drugs are of concern, because some consumers may be sensitive to the residues of such treatments. The use of antibiotics, many of which are the same as those used in human medicine, risks the emergence of antibiotic-resistant strains. Nevertheless, it is a sound principle that foods should not contain these materials, and the drugs must be subject to regulatory control. Countries with significant aquaculture industries will have requirements for the use of veterinary drugs in aquaculture as described earlier, although they might not be comprehensively enforced in some countries.

The risks (that is, the probabilities that hazards will result in food poisoning) may be less for aquaculture products compared with their wild counterparts. For example, the risk of contracting anisakiasis from farmed salmon, and perhaps from other intensively farmed marine vertebrate fish, is very low because of the use of pelleted feed. The risk of chronic poisoning from chemical toxicants, like heavy metals and highly chlorinated hydrocarbons, may also be lower in the case of farmed fish compared with catches from the wild because, generally, cultured fish are harvested at a relatively young age, when body burdens of the toxicants would be low. Furthermore, controlled culture conditions available to fish farmers have the potential to reduce the risk posed by known hazards.

However, good quality assurance practices are required by both feed suppliers and fish farmers to ensure that feeds and other inputs, such as water supplies, are not contaminated by harmful chemicals. Although controlled culture conditions may represent an advantage over wild capture fisheries in reducing public health hazards, there are no epidemiological data to indicate a difference in food safety risk between wild and cultured products. From the viewpoint of the food processing sector, there are many advantages from products from aquaculture compared with those from capture fisheries. For instance, products of uniform size or age are a real advantage for automation in fish processing. Such products tend to be less stressed and suffer from less physical damage compared with products that have been hauled in a net along the sea bottom or trapped in gill nets for extended periods. Thus there is a potential for farmed products to be of better physical and biochemical quality, enhancing the postharvest storage stability. Aquaculture products will be fresher than capture fishery products, because harvests can be timed to correspond with processing schedules and market demands. Handling, icing, and transport are all easier to carry out on land compared with conditions onboard fishing vessels on the high seas and the long distances that some fishermen must travel to reach fishing grounds.

The following review and discussion illustrates the current debates on aquaculture issues. Public health hazards of cultured products are influenced in the main by the environment in which the fish are cultured, the species under cultivation, and postharvest handling and processing.

Data on aquaculture production discussed in earlier sections of this chapter indicate the considerable contribution of nonindustrialized countries to supplies of aquaculture products. Furthermore, aquaculture in nonindustrialized countries with tropical and warm temperate climates accounts for the majority of this production. This relative dominance of freshwater aquaculture in nonindustrialized countries compared with industrialized countries is important in the context of public health. However, it is important to note that the majority of freshwater aquaculture products from nonindustrialized countries are for local consumption and not export purposes.

Although food poisoning is of concern in all countries, summary data on foodborne diseases for nonindustrialized countries are generally of poor quality. Only a few countries have effective mechanisms and systems for surveillance and reporting of food poisoning cases (Bryan et al. 1997; Guzewich et al. 1997; Notermans and Borgdorff 1997). Even in countries with comprehensive foodborne illness outbreak surveillance systems, it is estimated that only about 10% of cases of food poisoning are reported to the authorities. The most comprehensive surveillance data comes from industrialized countries, e.g., the United States, the United Kingdom, etc. (Bryan 1980; Bean and Griffin 1990; Cooke 1990; Todd 1990, 1992; Ahmed 1991).

Ahmed (1991) provides a detailed and comprehensive review of microbiological hazards of fishery products. Other recent accounts are found in Gibson (1992), Huss (1994), and Inglis et al. (1993). This literature is written in the context of industrialized countries. The literature addresses the hazards and public health aspects of fish and fishery products and focuses on the hazards of wild marine fish harvested from temperate waters that are processed and marketed in industrialized countries. Compared with temperate marine species, information on the public health hazards and risks of freshwater and marine species harvested from warm temperate and tropical waters is sparse. The Joint FAO/NACA/World Health Organization (WHO) Study Group reported on food safety issues associated with aquaculture (WHO 1999), and Howgate (1998) reviewed public health aspects of aquacultured products.

Biological Hazards of Aquaculture Products

Bivalve Mollusks Bivalve shellfish concentrate bacteria and virus particles from ambient water, which may result in the presence of pathogenic bacteria and viruses that constitute a public health hazard (Guzewich and Morse 1986). Shellfish should be safe for consumption after being cooked or heat processed; however, the risk of illness is considerably increased when they are eaten raw or after light cooking. Codes of practice for heat processing shellfish are based on the survival of bacteria, whereas viruses in bivalve shellfish can withstand higher heat processing temperatures than pathogenic bacteria (Abad et al. 1997) and therefore pose a greater risk of causing illness than bacteria. In some nonindustrialized countries, particularly those that have a large export market, official control of the safety of bivalves is exercised by a combination of regulations of the water quality from which they are harvested and by a requirement for depuration of harvested shellfish if the waters do not meet the required standards. These controls have proved inadequate to ensure food safety from viral illnesses, and many incidences of food poisoning are still attributed to wild-caught shellfish harvested from approved waters or those that have undergone depuration. (Gerba 1988). Numbers of enteric bacteria, the criterion for approval of waters, are not reliable indicators of the presence of pathogenic viruses. In addition, viruses depurate slowly from shellfish, and depuration times adequate to remove bacteria are not sufficient for viruses (Vaughn and Landry 1984; Beril et al. 1996; Abad et al. 1997).

Of the total world production of bivalve mollusks (13.1 million mt in 1998), 9.1 million mt, or 70%, come from aquaculture. Therefore, it is important to understand the factors that influence the degree of risk posed by cultured shellfish. Factors that influence the risk of food poisoning from bivalves are the fact that bivalves siphon vast amounts of water and, therefore, the microbiological quality of the growing waters, and the presence of harmful marine algae must be determined through rigorous shellfish safety control programs. However, there is no indication that culture practices employed for bivalves when grown in accordance with the requirements of a competent shellfish safety control program (as is required for wild harvest species) increase the risk of food poisoning (i.e., pathogens, viruses, biotoxins) above that of the risk posed by bivalves harvested from the wild. Bivalve aquaculture involves little intervention from the culturist other than to ensure that the culture waters and associated enforcement activities comply with the shellfish safety control program requirements. In addition, no supplementary food is given, drugs or other chemicals are not used, and there are no significant differences in growing conditions from those of wild crops.

Aquaculture of Vertebrate Fish and Crustacea in Marine and Brackish Water Environments

Pathogens Several species of bacteria pathogenic to humans are indigenous to the marine environment (Ahmed 1991; Ward and Hackney 1991; Gibson 1992; Huss et al. 1997; Dalsgaard 1998; Dombroski et al. 1999). These bacteria include *Vibrio* spp., *Clostridium botulinum*, *Listeria monocytogenes*, *Aeromonas hydrophila*, and *Plesiomonas shigelloides*. Of these, *Vibrio* spp. are the most significant for human health. *Vibrio parahaemolyticus* is well recognized as a cause of food poisoning from seafoods, although other species, that is, *V. cholerae* and *V. vulnificus*, have also been implicated in outbreaks. Marine fish in estuaries or inshore waters receiving sewage pollution may also become contaminated with enteric organisms. These organisms can include pathogens such as *Salmonella* spp., *Shigella* spp., and pathogenic strains of *Escherichia coli*. In many instances where fish and crustacean shellfish have been implicated in food poisoning by enteric organisms, the infective dose of the pathogenic bacteria occurred during postharvest handling and processing of the product, rather than being present in the fish or shellfish when harvested.

Where marine fish from inshore or estuarine waters are contaminated with sewage-borne human enteric organisms, it is likely that the incidence of contamination will be low. Careful removal of the gut, so that the muscle tissues are not contaminated, and cooking further reduce risks to human health. There is no reason to expect that the risk of food poisoning from farmed marine fish, or from products prepared from farmed marine fish, is any higher than that from the corresponding species in the wild state. In fact, the risk from fecal organisms might be lower, because fish species farmed in the marine environment are usually not very tolerant of polluted waters, and therefore farms are typically situated in clean waters.

Shrimp production is an important aquaculture industry in several nonindustrialized countries. Shrimp are commonly cultivated in brackish water coastal areas, and pathogenic organisms similar to those present in the marine environment are likely to be present. The coastal siting of shrimp farms may place them at a greater risk of contamination from fecal microorganisms and viruses because they are likely to be closer to the source of contamination. Additionally, animal excreta are often used as fertilizers. For example, Reilly and Twiddy (1992) isolated *Salmonella* from 18% of samples of shrimp from 131 farms in one of the Southeast Asia producing countries. Some of these farms used chicken manure as fertilizer, but *Salmonella* was also isolated in shrimp from farms using only pelleted feeds. There was a weak relationship between the incidence of isolation of the organism and type of feeding. Bhaskar et al. (1995) and Bhaskar et al. (1998) reported the presence of food-poisoning organisms in farmed shrimp and in the pond environment.

Reilly et al. (1992) reviewed the occurrence of *Salmonella* in cultured tropical shrimp and concluded that *Salmonella* spp. should be considered as a component of the natural microbiological flora of farmed shrimp. They also commented on the use of chicken manure as a fertilizer (see also Dalsgaard and Olsen 1995) and referred to other sources of contamination such as birds and animals and drainage from the land. Food-poisoning organisms found on farmed shrimp are readily destroyed by heat and, if contaminated, should be safe to eat after normal cooking and handling procedures. There have been reports of incidents of food poisoning from consumption of cooked and peeled shrimp, but these incidents are most likely due to cross-contamination, poor handling procedures, and poor storage conditions of the shrimp after cooking.

Parasites Nematode parasites are commonly present in various tissues of fish caught in the wild. They are most frequently found in the liver and belly cavity but are also present in the flesh (Deardorff and Overstreet 1991). The most common species of nematode causing disease in humans is *Anisakis simplex*, hence the name of the disease—anisakiasis (Sakanar and McKerrow 1989). The other species involved in anisakiasis in North America, Europe, and Japan is *Pseudoterranova decipiens*. These nematodes occur worldwide and are undoubtedly present in fish in marine waters of nonindustrialized countries, although the documentation on the prevalence in those countries is much less extensive than that for industrialized countries. The definitive hosts of the parasites are piscivorous marine mammals such as seals, with intermediate hosts being marine invertebrates and fish. Although the incidence of nematode infestation of fish is high, anisakiasis is not a common disease, because the parasite is killed by normal cooking and by freezing. There is a significant risk of disease from fishery products consumed raw. The incidence is also high after only mild processing of fish, such as salting at low concentrations or cold smoking. Many countries now require that fish used for mildly processed products be frozen before processing if they are intended to be consumed without cooking.

The parasite infects the fish through its diet. Thus farmed fish that are fed processed, pelleted feeds should not become infected. However, it must not be assumed that farmed marine fish are necessarily free of nematode worms, because in some systems the cultured fish are fed with raw parasitized fish that may allow the farmed product to become infected.

To control the parasite hazard, trash fish and offal used for fish feed should be heated or frozen to a temperature sufficient to render any biological agents noninfectious. Presently, compared with freshwater aquaculture, the production of finfish by mariculture is small. Nevertheless, infestation or contamination of product with nematodes in feeds may be more of an issue for nonindustrialized than industrialized countries because of the increased costs associated with pelleted feeds compared with the use of trash fish.

Aquaculture in Freshwater

Pathogenic bacteria found in marine environments can also be found in freshwater, although the composition of the flora is not the same.

Human pathogens, *Edwardsiella tarda*, *Aeromonas hydrophila*, and *Pseudomonas shigelloides*, are more common in freshwater environments than in marine environments and have been associated with farmed fish.

Land-based aquaculture systems and inland water bodies in which fish are cultivated may pose a much higher risk of contamination with enteric organisms compared with offshore marine culture systems, because water supplies to these culture systems are more likely to contain fecal material from animals, birds, and humans. However, there are a few reports concerning the presence of enteric pathogens in farmed fish cultivated in unfertilized systems (Wyatt et al. 1979; Saheki et al. 1989; Nedoluha and Westhoff 1993). These reports are from industrialized countries that would be expected to have strict controls over the disposal of sewage. Reports concerning pathogens in unfertilized aquaculture systems in tropical countries, without effective systems for disposal of sewage, are sparse. Ogbondeminu (1993a) compared the incidence of enteric bacteria in fish and water from three aquaculture sites in the Kanji Lake basin, Nigeria. *Salmonella* was not isolated from the guts of fish cultured in unfertilized concrete tanks fed by reservoir water, but it was isolated at the two sites fertilized by wastewater or chicken manure.

Pathogens The discussion presented in the previous section indicates a low incidence of enteric pathogens in fish cultured in unfertilized ponds. Thus the bacteriological public health hazards would not be expected to be different in character and intensity from those in fish caught from the wild. However, most data come from countries with temperate climates. It is unfortunate that data from tropical regions are lacking.

Another potential hazard is the widespread practice of using human and animal excreta as fertilizers in pond aquaculture and raising fish in wastewaters. A high proportion of the world's aquaculture production of vertebrate fish is raised in these systems. In a report outlining the safe use of wastewater and excreta in agriculture and aquaculture (WHO 1989a), it was estimated that as much as two-thirds of the cultured vertebrate fish may be produced in ponds fertilized with human or animal excreta. The WHO (1989a) Guidelines refer largely to the use of wastewater in aquaculture in Calcutta, India. However, the practice is much more widespread, if only on an experimental basis (Payne 1984; Oláh et al. 1986; Edwards and Pullin 1990; Prein 1990; Ninawe 1994; Strauss 1996).

A number of reviews and reports have described the use of excreta in aquaculture and the raising of fish in sewage wastewaters. Most reports have identified health hazards associated with these practices (Bryan 1977; Edwards 1980; Payne 1984; IRCWD 1985, 1988, 1990; Sharma and Oláh 1986; Sin and Chiu 1987; WHO 1989a; Edwards and Pullin 1990; Pillay 1992; Strauss 1996). Nevertheless, the use of human excreta is less widespread compared with animal excreta, because there are implicit or explicit taboos against its use (Cross 1985; Furedy 1990). Furedy (1990) comments that in China the use of human excreta in aquaculture is being phased out.

Typically, in lagoon-based wastewater treatment plants, sewage passes through a series of connected ponds. Solids that are unsuitable for raising fish settle in the first and second ponds with fish being grown in the third and subsequent ponds (Edwards and Pullin 1990). There are a few reports of the microbiology of fish grown in sewage effluents (Buras et al. 1985, 1987; Slabbert et al. 1989; Balasubranian et al. 1992; Ogbondeminu 1993a, 1993b; Bhowmik et al. 1994; van den Heever and Frey 1994; Eves et al. 1995). This suggests that fish can be cultured in wastewater treatment ponds without their posing a significant risk to public health, as long as some safeguards are met (Edwards 1990; Buras 1993). For example, it would not be advisable to raise fish for human consumption in ponds early in the sequence of treatment (WHO 1989a). Strauss (1985) has reviewed pathogen survival in wastewater systems and concludes that a system with three or more ponds giving a residence time of at least 20 days will give three log reductions of fecal coliforms and *Salmonella* spp. Fish raised in ponds late in the sequence of treatment might be passive carriers of enteric pathogens on the skin and in the gut, but judging from data on the microbial flora of fish from natural waters, the risk to public health might not be greater than that posed by wild fish harvested from nearshore environments.

Although the risks to human health from bacterial food poisoning from fish raised in wastewater or in ponds fertilized by animal or human excreta might appear obvious, there are no reports of food poisoning incidents attributable to these sources. However, there are suggestions from epidemiological studies in communities consuming fish from wastewater aquaculture or manured ponds of increased incidences of diarrheal illnesses (Blum and Feachem 1985). The topic obviously warrants further study for better understanding of the hazards associated with aquaculture. The dearth of reports of food poisoning incidents containing aquaculture in wastewater or excreta-fertilized ponds should be considered to assist in the development of management practices that limit public health risks.

Parasites Well-attested (although perhaps not all that well recognized outside of Asia) public health hazards relate to the consumption of freshwater vertebrate fish and crustacean shellfish from warm, temperate, and tropical climates that may be infected with trematode parasites. The public health significance of foodborne trematode infections has been reviewed (WHO 1995), and it was estimated that somewhere in the region of 40 million people worldwide are affected by fishborne trematode infections. This prevalence is far greater than the numbers affected by bacterial food poisoning from fish. The most important parasites, based on the number of people affected, are species of the genera *Clonorchis*, *Opisthorchis*, and *Paragonimus*. All have similar life cycles involving a definitive host and two intermediate hosts. The definitive host is humans and other mammals, the first intermediate host is a snail, and the second intermediate host is either a freshwater fish or a crustacean shellfish. Humans and animals are infected by eating raw or minimally processed fish or shellfish infested with the parasites; these parasites are highly infective. Again, proper cooking renders the parasite harmless.

Howgate (1998) has reviewed the occurrence of trematodes in freshwater fish and the implications for aquaculture. Publications by Cross (1991) and the Institute for Tropical Endemic Diseases (1984) provide extensive information on trematode parasitic infections in Asia and Southeast Asia, including data that are relevant to diseases transmitted by freshwater fish and crustacean shellfish. A useful summary of the conclusions of the second of these proceedings is provided by Brier (1992). Rim (1982) presents a review on clonorchiasis, the disease caused by *Clonorchis* spp., and associated public health implications. Malek (1980) gives a detailed account of snail-transmitted parasites, including those listed in the previous paragraph, and the diseases they cause. In addition to these reviews and conference proceedings, individual papers in journals give information on endemic areas and the incidence of infection in fish (examples are Vicharsi et al. 1982; Ditrich et al. 1990; Park et al. 1993; Yu et al. 1994a; Yu et al. 1994b).

Where trematode diseases are endemic, and there is no evidence to the contrary, it should be expected that freshwater fish cultured in systems receiving fecally contaminated surface water will be infected with trematode parasites. Kino et al. (1998) reported on the epidemiology of clonorchiasis in an area in Vietnam and the associated infestation of secondary hosts of the causative parasite. They found a higher incidence of *Clonorchis cercariae* in snails from aquaculture ponds than in rivers and rice fields, with metacercariae being found in 100% of larger specimens and 54% of smaller specimens of silver carp (*Hypophthalmichthys molitrix*) being cultured in the ponds. This high incidence is perhaps to be expected; cercariae shed by the first intermediate hosts (snails) would have a higher probability of finding a secondary host (fish) in the crowded conditions of a fishpond than in more dispersed wild fish populations. Because the eggs of cestode parasites are shed in the feces of infected animals and humans, this may lead to higher incidences of trematode infection in excreta-fertilized systems, with a consequent higher risk of infection in consumers.

The WHO report on foodborne trematode infections (WHO 1995) suggests a link between excreta reuse in aquaculture and trematode infections but does not provide any specific information or data. Naegel (1990), in a review of health problems relating to the use of animal excreta in aquaculture, writes that there is potential for the spread of trematodes by such practices but does not provide any examples. Cross (1984) attributed an increase in the incidence of clonorchiasis in Hong Kong to the importation of pond-reared fish from China. Chen (1991) reported that clonorchiasis was common in fish farms in parts of Taiwan and associated the high incidence of infection with the use of pig manures as fertilizer. However, there is a lack of extensive data, not only on the relative rates of infestation of aquaculture products and of products from the wild, but also on the effect of cultural practices on risks to human health.

Freshwater vertebrate fish culture contributes a large portion of the world's total aquaculture production, with much of it being produced in countries in which trematode infections are prevalent. Control over environmental conditions may potentially reduce, if not eliminate, the hazard of trematode parasites in farmed freshwater fish. Lima dos Santos (1997) and Reilly and Käferstein (1997) have discussed the application of HACCP principles for the control of parasite hazards. However, to date HACCP has been applied only to established processing systems with no methodology for the control of parasite hazards in farmed fish being developed, much less any extensive studies being completed and published. In countries where fishborne trematode infections have been brought under control, the control mechanisms have been mainly based on case detection and treatment, public education, and improved sanitation (WHO 1995). This breaks the chain of infection as far as human hosts are concerned but could leave reservoirs of infection in domestic and wild animals. Therefore, restrictions on the access of animals to fish farms may be required to eliminate the public health risk.

Specific practices for the control of parasite hazards at fish farms have been suggested and should be considered. Discontinuing the practice of direct fertilization of ponds with human or animal excreta and the treatment of excreta by composting before use (Strauss 1985) could prevent primary infection of the water. An obvious strategy for control of trematodes in aquaculture would be to eliminate the mollusk host from the system (Agarwal and Singh 1988), which has proved effective in schistosomiasis control (McCullough 1990). The population of snails in a pond can be eliminated by the use of molluscicides and by fallowing; snails must be prevented from recolonizing the ponds. Such an approach must be integrated with a system that includes controls on what is introduced into the system to prevent introduction of snail and trematode eggs. However, this approach could be difficult to apply in small-scale fish farming. Introduction of molluscivorous fish species to the pond could be a further control measure for snails. This strategy, by itself, has not proven to be completely effective in Africa (Slootweg et al. 1994) but could be a useful component in an integrated approach to snail control. Polyculture systems often include species that are molluscivores. It would be useful to know whether snail populations and the degree of infestation of fish by trematodes in polyculture systems are affected by the presence of these species.

Another approach in controlling parasite hazards might be to treat the fish to kill the metacercariae (Hirazawa et al. 1988; Mitchell 1995). However, it would be unwise to use praziquantel (a drug that has proven very effective in treating parasite diseases in humans) in such programs, because its widespread use and misuse might induce resistance of the drug in the target parasite. This drug should be reserved for human therapeutic use. Other drugs are available (Hirazawa et al. 1988) for treating trematode infections in fish, including natural products that could satisfy the requirements of organic production. Fish targeted for chemical treatment of parasites would most likely need to be transferred from growing ponds to tanks for treatment, followed by a withdrawal period for depuration.

Chemical Hazards

Bodies of water often become the ultimate recipients of a very wide range of chemicals, some derived from natural processes and others originating from human activities (xenobiotics). There is a very large selection of literature (including several reviews) on chemicals and chemical pollutants in the marine environment. A smaller selection of literature relating to the freshwater environment has been published. There does not appear to be a review of the human health hazards of chemicals and pollutants in the context of aquaculture other than addressing veterinary drugs and chemicals used in aquaculture. The latter is discussed in the **ENVIRONMENTAL HEALTH ISSUES** section of this chapter.

Metals Elevated concentration of metals in the edible portions of aquaculture products is relevant to public health rather than the concentrations that are present in the culture water. Metals enter the fish by absorption through the gills or by absorption from food, but it appears that the latter is the more important route (Biddinger and Gloss 1984; Dallinger et al. 1987). The concentration of metals accumulated in tissues is usually greater than those in water or the food. However, the degree of bioaccumulation differs among metals, fish species, and tissue types. Metal concentrations are lowest in the muscle tissue of vertebrate fish and tend to increase in parts of the fish that are usually removed from larger fish before consumption (i.e., kidney and liver).

Vertebrate fish can regulate the concentrations of inorganic forms of metals in muscle tissue. Typically, concentrations of metals (other than possibly mercury) do not exceed regulatory or recommended limits, even when the fish are harvested from metal-contaminated lakes or ponds or from marine environments exposed to metal contamination. Sewage often contains high concentrations of heavy metals, but measurements in farmed fish, even those grown in sewage-fertilized systems, show that the contents of metals (with the possible exception of mercury) are below regulatory or recommended limits (Tarifeño-Silva et al. 1982; Turner et al. 1986; Sin and Chiu 1987; Ackefors et al. 1990; Guerrin et al. 1990; Prein 1990; van den Heever and Frey 1994; Pandey et al. 1995; Rajan et al. 1995).

Although there is a large database of mercury concentrations in animals from capture fisheries, there are few reported measurements for aquacultured products (Turner et al. 1986; Guerrin et al. 1990; De Boer and Pieters 1991). In all cases, the aquaculture samples had mercury contents of less than 0.5 mg/kg, used by most regulatory agencies as the permitted maximum limit in seafood for human consumption.

Methylmercury, mercury bound in an organic complex, is a significant exception to the regulation of metals in muscle tissue by vertebrate fish. There is a substantial amount of literature on this subject.

Both the Group of Experts on the Scientific Aspects of Marine Pollution (GESAMP) (1986) and WHO (1989b) have conducted reviews of aquatic environmental and human health aspects. Inorganic mercury can be methylated by biological (predominantly microbiological) processes in the aquatic environment. More than 95% of the total mercury in edible portions of fish and invertebrate tissue is in the form of methylmercury (Bloom 1992). Methylmercury bioaccumulates at higher trophic levels so that the highest concentrations are found in predatory fish.

Mercury accumulates in fish over its lifetime, thus mercury body burdens in fish in a particular environment are greater in the older, and hence, larger fish. Farmed fish are usually harvested at a young age and would be expected to have a low body burden. Uptake of mercury is influenced by the chemistry of the ambient water. Water in ponds used for aquaculture commonly has a high pH, elevated hardness levels, and a high organic content, conditions that inhibit mercury uptake (Håkanson 1980).

There is clearly a need for more information regarding the mercury content of products from aquaculture and the potential influence of management practices on reducing possible public health risks. However, the risk of inducing mercury toxicosis through consumption of vertebrate species produced through aquaculture may be lower than that associated with corresponding species caught from the wild. Methylmercury is taken up predominantly from ingested food, and where fish are fed formulated diets and the feeds have low mercury contents, harvested products will have low concentrations of mercury in their tissues (De Boer and Pieters 1991).

Other metals and metalloids of concern for human health exist in a number of forms and valence states. The chemistry of their fates in the aqueous environment is complex. The pH of the water plays a large role, because metal solubility decreases with increasing pH. Freshwater tends to be alkaline, and aquaculture pond systems are usually maintained with a pH above 8. Ponds usually have anaerobic, organic-rich sediment. Under these conditions, metals tend to precipitate in the sediment as insoluble sulfides or hydrated oxides (Wiener and Giesy 1978).

Tin compounds are often used in antifouling paint for ships, but the use of tin tributyl and triphenyl compounds as antifouling treatments for nets or pens in aquaculture has generally been banned throughout the world. Use of such paints has raised concerns because of toxic effects toward aquatic invertebrate organisms, in particular, gastropod mollusks (Hall and Pinkey 1985). WHO has published a monograph on tributyltin compounds, which includes a review of possible human health hazards. Tributyltins have been used in the past, at least on a trial basis, for treating nets and pens used in mariculture to prevent fouling. It has been shown that salmon and other species held in treated nets accumulate tin in their tissues (Short and Thrower 1986; Davies and McKie 1987; Amodio-Cocchieri et al. 2000). Pullin, in the discussion following a paper in a conference (Pullin et al. 1993), referred to the use of organotins as molluscicides in rice farming. This gives reason for concern because aquaculture is often integrated with agriculture, particularly in rice-growing areas, and irrigation water from rice cultivation could find its way into land-based and coastal aquaculture systems.

The WHO monograph reported that the working group compiling the report was unable to quantify the risk to humans posed by the consumption of fish and shellfish contaminated by tributyltin, but Schweinfurth and Günzel (1987) extrapolated from animal studies to suggest an allowable daily intake (ADI) of 3.2g/kg body weight. For a 70-kg adult, this amounts to an ADI of 0.22mg. The maximum concentration of tributyltin found by Davies and McKie (1987) in salmon was 3.05mg/kg, which equates to an intake of 0.46mg, assuming consumption of a 150g portion of salmon muscle. It would be uncommon for a person to consume salmon daily, but the intake calculated is of the same order as Schweinfurth and Günzel's proposal.

Heavy metals, and some metalloids such as arsenic, are mostly present in the aqueous environment as a result of geochemical processes that mobilize them into solutions so that they ultimately enter watercourses and other bodies of water. Anthropogenic sources include mining, metalworking, and industrial processes. Municipal sewage, unless it comes entirely from domestic sources, often contains a range of metal salts. The origins, fates in the aqueous environment, and human health hazards of some metals have been reviewed in GESAMP publications (GESAMP 1985; 1986). Although these articles were written from the perspective of marine pollution, much of their contents and conclusions are valid for the freshwater environment. Nevertheless, concentrations of metals are low in the open oceans and in unpolluted waters. Coastal areas, which are often favored sites for mariculture, may contain high concentrations of metals as a result of transport by rivers and from anthropogenic sources (GESAMP 1987).

Industrial Chemicals and Agrochemicals Considerable literature exists on the presence and fate of highly chlorinated hydrocarbons in wild aquatic species. A limited but growing literature exists on highly chlorinated hydrocarbons in aquaculture systems or aquaculture products (Perry et al. 1983; Vojinovic-Miloradov et al. 1992; Lehmann 1993). These reports show that, although fish contained highly chlorinated hydrocarbons, concentrations were below the maximum levels generally permitted in foodstuffs. Sewage and wastewaters often contain high levels of industrial chemicals. Surface runoff waters may contain high concentrations as well, but Turner et al. (1986) analyzed fish and shrimp grown in domestic sewage in South Africa and found only very low levels of chlorinated insecticides.

Pullin et al. (1993) reported that there were no data on polychlorinated dibenzo-*p*-dioxins (PCDD) or polychlorinated dibenzo-*p*-difurans (PCDF) in farmed aquatic products. Ingestion from food is considered the main route of uptake and transfer of PCDD (Lieb et al. 1974; Hansen 1980; Fisher et al. 1986; Tatem 1986; de Boer and Pieters 1991; Jackson and Schindler 1996). Organochlorine contaminant concentrations in fish raised in intensive aquaculture systems, in which fish are fed a formulated diet, will be low unless the feed itself contains these compounds in more than trace amounts. Highly chlorinated hydrocarbons are pervasive in the environment, and all feeds, including vegetable and animal, will contain some low levels. In addition, concentrations in the feed are decreasing with time now that production of polychlorinated biphenyls (PCB) has been curtailed and controls over emission of dioxins have been introduced. Hilton et al. (1983) fed rainbow trout for 24 weeks on diets containing fish meals with different levels of highly chlorinated hydrocarbons and showed that the trout accumulated the contaminants in proportion to the amounts in the diets. Furthermore, Makarewicz et al. (1993) showed that fish could be cultivated in polluted water and yet have low concentrations of chlorinated pollutants, provided the diet was low in contaminants.

Unfortunately, data on chemical contaminants in fishery products from Asia and Southeast Asia, where most freshwater aquaculture in nonindustrialized countries is undertaken, are sparse. The flux of organic contaminants in aquatic ecosystems, their distribution among different compartments of the system, and their bioaccumulation at different trophic levels can be modeled, and these models have been successfully applied to field situations (Barber et al. 1991; Mackay 1991; Borgmann and Whittle 1992; Gobas and McCorquodale 1992; Abbott et al. 1995; Cole et al. 1999). It would be useful to apply these models to some representative freshwater aquaculture systems to predict how persistent highly chlorinated hydrocarbons would transfer to, and distribute among, trophic compartments in aquaculture pond systems.

In Asia, most of the aquaculture production comes from inland sites that are associated with agricultural activities and must be considered at risk from contamination with agricultural chemicals. Pesticides like DDT and lindane were extensively used in Asian countries to control pests in agriculture and mosquito vectors carrying malaria. Although the use of chlorinated insecticides has been, or is being, phased out in these countries, highly chlorinated hydrocarbons are very persistent in soils and can leach into surface waters.

Aquaculture systems can be impacted by acute and/or chronic discharges of organic pollutants. Acute pollution results from single, short-lived discharges, such as accidental spillages from chemical works into watercourses or as a result of vessel groundings. In the case of onshore systems, including integrated aquacultural/agricultural systems, pollution of the aqueous environment can arise from episodic agricultural treatments such as crop spraying. In acute instances, aquaculture systems at risk should be readily apparent and the operator may be able to avoid or mitigate the effects. Additionally, regulatory authorities might intervene to prevent distribution of contaminated stock until the danger has passed and the fish or shellfish are purged of the contaminant. More difficult to control is chronic contamination. Main routes of chronic contamination in aquaculture systems include the use of polluted water supplies, leaching of agricultural or industrial chemicals from treated or contaminated soils into surface waters, and deposition from the atmosphere. Most industrial and agricultural chemicals are readily degraded by chemical and biological processes in soils and waters, do not bioaccumulate to any large extent, and are rapidly eliminated from fish (Tsuda et al. 1994). Of more concern are highly chlorinated hydrocarbons, especially from the following three groups in particular: chlorinated insecticides such as DDT, dieldrin, lindane, and their degradation products; PCB; and PCDD and PCDF (Ackefors et al. 1990; Phillips 1993, 1995; Delzell et al. 1994; Wells and de Boer 1999).

ENVIRONMENTAL HEALTH ISSUES

The following sections list and discuss many of the controversial, ongoing environmental issues associated with global aquaculture operations in nonindustrialized countries. The expectation is that this information will help elicit discussions that will lead to new approaches to address and resolve the problems.

Introduction

Although there are various definitions of the term "environmental health," it is interpreted here to be primarily concerned with the factors that have a negative impact on the environment and that may lead to a disruption in the supply of environmental goods and services that may subsequently lead to negative impacts on human and animal health.

Aquaculture and the Environment

Aquaculture relies on a wide range of natural resources or environmental goods. These include land (or space in larger water bodies and the sea) to site an aquaculture facility; materials (soil, timber, stone, steel, etc.) for construction of the system; infrastructure ancillary facilities (roads, offices, etc.); water containing oxygen for respiration and to support the animals and flush away waste; and feed and fertilizers to enhance production and seed (eggs, larvae, postlarvae, or fry) for stocking. When production is carried out in ecologically open systems, wastes are produced and discharged into the environment. These waste products may include uneaten food; fecal, urinary, and excretory products; chemical agents used to condition sediments and culture water and combat disease (chemotherapeutants); feral (escaped) culture organisms; and pathogens. After the discharge of waste products, aquaculture is dependent on essential environmental services, i.e., the replenishment of oxygen and dispersal and assimilation of nutrients, to ameliorate possible negative environmental impacts that could result in negative feedback on growth and production in the culture system.

Resource Use

The following examples illustrate the need for improved selection criteria to ensure proper siting of aquaculture operations to minimize negative impacts. Stakeholders have identified negative social and environmental impacts associated with the development of large-scale commercial aquaculture systems in mangrove areas. An important social impact associated with mangrove conversion is the exclusion of marginal groups who previously relied on the natural productivity of the mangrove (Nickerson 1999). From an environmental perspective, mangrove habitats sustain biodiversity, regulate the accretion of sediments, influence local hydrological regimes, and convert nutrients from terrestrial ecosystems to organic inputs supporting near-shore detrital food webs. These environmental functions and processes associated with mangroves are lost through conversion to aquaculture ponds. Primavera (1991) reviewed the ecological, social, and economic impact of conversion of mangroves to aquaculture production systems. Although much attention has been given to the conversion of mangroves in Asia to guard against loss, similar consequences may accompany the proposed conversion of mangroves for aquaculture in other regions, e.g., eastern Africa (Semesi 1998).

Water Appropriation of water for aquaculture and the subsequent discharge of wastewater can have a significant impact on the local hydrology as well as chemical composition of water and sediments in the receiving environment and should be addressed through improved site selection and wastewater management. Major physical impacts can include the modification of flow regimes and hydrological conditions in the receiving environment and the effect of sedimentation. Changes in the chemistry of the receiving environment are largely related to the release of nutrients and chemical agents used to treat disease. However, oxygen consumption within the aquaculture facility and the biological oxygen demand associated with discharged waste also impact water quality.

In addition, deposition of particulate matter entrained in aquaculture wastewater can be unsightly. The increased siltation can smother invertebrates and macrophytes, increasing substrate embeddedness, reducing interstitial water flow, and restricting the access of certain fish species to substrates suitable for forming spawning grounds. Sedimentation is more likely to occur when the dilution capacity of the receiving environment is limited (Jones 1990). Furthermore, particular habitats are more vulnerable to siltation, for example, deeper reaches of rivers downstream of aquaculture operations and directly beneath cage farms.

Seed This example illustrates the need for the development of new hatcheries and brood stock rearing facilities to address the growing requirement for good quality seed by aquaculture. The culture of numerous aquatic species depends on seed, fry, larvae, postlarvae, or gravid females collected from wild populations. Carp culture in many Asian countries has relied on wild fry captured from rivers; however, returns from these fisheries have diminished and hatchery production now supplies much of the demand for carp fry. Although some of this decline in the capture of wild fry may be attributed to the collection of seed for aquaculture, other factors (e.g., intense fishing pressure on mature fish stocks, migratory routes made impassable by flood protection schemes, and increased pesticide use) have also contributed to this decline. Collecting fry to supply the shrimp aquaculture sector in many Asian countries presented a valuable employment opportunity for many poor people in coastal communities. Nevertheless, this intense fishing pressure has been linked to declining stocks. Furthermore, the harvesting strategies employed have resulted in a significant by-catch that is discarded, which in turn results in a decline of the number of other species in near-shore areas. Where the abundance of shrimp fry has decreased, artisanal fishing methods have been replaced with more commercial strategies. For example, traditional push nets used by individuals to collect fry in shallow water now compete with offshore shrimp fisheries, which use boats and nets to capture gravid females. Although the returns from collecting gravid females may be higher, so are the potential impacts on shrimp stocks. Harvesting mature females may significantly reduce recruitment in subsequent years.

Feed Parallels may be drawn between the role of wild seed and feed components derived from the local environment. In many instances, operators of small-scale aquaculture production systems initially depend on feeds from plants and animals collected from the wild. Aquatic plants (e.g., duckweed) are commonly harvested from natural water bodies to feed herbivorous fish stocked in rural ponds. Natural populations of invertebrate species (e.g., snails) are also used to provide feed inputs for aquaculture, most notably the *gher* farming systems used to culture prawns in Bangladesh. This practice has led to the loss of snail fauna in large parts of the country. The development of alternative feeds based on locally available by-products from agricultural and food processing would help reduce pressure on native flora or fauna.

Wastes and Environmental Services

Static water pond-based aquaculture systems, which account for the majority of freshwater aquaculture production in nonindustrialized countries, may be considered relatively closed to the outside environment. Inputs include seed, feed, and fertilizer introduced to the pond. Outputs from the farming system consist mainly of harvested products, whereas waste products are largely assimilated within the pond ecosystem (Edwards 1993; Berg et al. 1996). However, where appreciable quantities of waste products are discharged from the aquaculture system, the potential for environmental degradation exists.

Eutrophication Eutrophication has been defined as "the enrichment of water by nutrients, especially compounds of nitrogen and phosphorus, causing an accelerated growth of algae and higher forms of plant life to produce an undesirable disturbance to the balance of organisms present in the water concerned" (Gowen 1994). Nutrients released from aquaculture have been associated with increased primary production in the receiving environment. Although nutrient inputs from aquaculture can be locally very important, several studies have shown that they contribute only a relatively small proportion to the overall anthropogenic input to aquatic ecosystems compared with other nutrient sources. It has been estimated, for example, that shrimp aquaculture in the coastal states of Mexico contributes 1.5% and 0.9% to the total input of nitrogen and phosphorus, respectively to the marine environment (Páez-Osuna et al. 1998). However, such studies on the extent of the problem in nonindustrialized countries are limited.

The role of aquaculture wastewater in eutrophication will depend on the nature of the receiving environment, including background nutrient status and water chemistry, hydrology, and carrying capacity. Nutrients released from aquaculture generally represent a relatively small proportion of the overall input to aquatic ecosystems. Waste fractions primarily arise from uneaten feed, excreta, fecal material, and chemical treatments used to maintain water quality and eradicate disease (Beveridge and Phillips 1993). Problems associated with oxygen consumption within aquaculture facilities and the subsequent compensation for biological and chemical oxygen demand can be compounded by secondary oxygen consumption (Bailey-Watts 1994). Decomposition of 1 g of fish feed has been estimated to consume 1.65 g of oxygen. However, phosphorus contained in the feed can sustain 10–40 g of plant production, which will consume 2–6 g of oxygen when decomposed.

Shifting Trophic Status and Interactions Increasing levels of sedimentation and organic enrichment beneath cages and rafts supporting shellfish frequently cause a decrease in benthic taxa diversity and an increase in the abundance of pollution-tolerant species (Karakassis et al. 1999; Stenton-Dozey et al. 1999). However, observable impacts typically do not extend much beyond 100 m from the culture facility (Costa-Pierce 1996). Environmental impacts from organic enrichment from aquaculture are dependent on several hydrological factors (i.e., flushing, resuspension, and dispersion). De Grave et al. (1998)

suggested that the highly dissipative nature of a site used to culture oysters (*Crassostrea gigas*) resulted in little organic enrichment or change to the surrounding benthic community.

Feral Culture Organisms Environmental impacts associated with the unintentional introduction of nonnative species from aquaculture facilities have been reviewed extensively (Welcomme 1988; Arthington and Bluhdorn 1996; Bardach 1997; Beardmore et al. 1997). Beveridge and Phillips (1993) proposed five distinct environmental impacts associated with the escape of species from aquaculture facilities, the first being the disruption that escapees can cause to the receiving environment. The widespread transfer of common carp (*Cyprinus carpio*) for aquaculture and establishment of recreational fisheries has resulted in wild populations becoming established in numerous countries (Arthington and Bluhdorn 1996).

The second category of environmental impacts ascribed to escapees by Beveridge and Phillips (1993) was disruption to the resident community, primarily through competition and predation. Organisms escaping from culture facilities may compete with resident communities for food and habitat. For example, Fernández and Fernández (1998) indicate that the introduction of rainbow trout to Tucuman province, Argentina, initially for sport but more recently for aquaculture, resulted in a decline in the diversity of benthic invertebrate communities through predation. Furthermore, competition for habitat with resident fish populations is believed to have displaced native species that were adapted to the winter droughts and summer floods that characterize streams in the region.

Beveridge and Phillips (1993) outlined the possible implications of escapees breeding with resident populations in their third category of impacts. The use of native species in farming can pose a threat to the integrity of strains adapted to local conditions. For example, native populations of *Clarias macrocephalus* in Thailand are at risk. Hybrids (F_1) produced by crossing the native species with the African walking catfish, *C. gariepinus*, could escape from culture facilities and breed with the resident species (Beardmore et al. 1997). This situation is repeated in Bangladesh, where it is feared that hybrid fish may escape from culture facilities derived from crossing *C. batrachus* and *C. gariepinus* and breed with local populations of *C. batrachus*.

Finally, the escape of transgenic organisms is a growing concern as the possibility of widespread culture increases (Beardmore et al. 1997). However, the possible transfer of genetic material from transgenic organisms to native populations is likely to depend on a number of factors, particularly the overall fitness of the organism (Arthington and Bluhdorn 1996).

An on-site aquaculture management plan is an essential prerequisite for the control of escaped cultured organisms into the environment.

Microorganisms The release of pathogenic organisms (e.g., viruses and bacteria) from aquaculture facilities has been reported in a number of studies (Arthington and Bluhdorn 1996). Nevertheless, the impact of introduced diseases originating from aquaculture facilities on wild populations has not been widely assessed (Pillay 1992). Epidemiological studies are being undertaken to elucidate linkages between disease outbreaks in wild and captive populations and aquaculture activities. Evidence from these studies points to a number of possible causative agents and contributing factors. Unfortunately, such studies are not routinely conducted in nonindustrialized countries because they are expensive and require considerable laboratory infrastructure.

Chemotherapeutants In general, the use of chemicals in temperate aquaculture is closely regulated to help ensure operator safety, protect the environment, and safeguard the quality of the product (see Chapter 3). However, in nonindustrialized countries where aquaculture production has increased dramatically, the rate of expansion has frequently overwhelmed attempts to regulate and monitor the industry. Consequently, the problems associated with the use of chemicals in aquaculture may be more acute in nonindustrialized countries as a result of a poorly defined regulatory framework. It can be expected that, as institutional arrangements evolve, the operators of aquaculture facilities will require appropriate management plans and treatment regimes to limit the release of chemicals in wastewater.

In many cases, a large proportion of medication is not taken up and retained by the culture organisms but passes out of the aquaculture system to the receiving environment (Weston 1996). Furthermore, antibacterial medications are routinely delivered as feed additives; however, diseased fish generally have a suppressed appetite, decreasing the proportion of medicated feed consumed. The discharge of therapeutants from fish farming operations is increased further as the majority of active agents ingested are not absorbed in the gut but expelled unmodified in feces. Thus more than 95% of some antibacterial agents administered in aquaculture are discharged from culture facilities (Weston 1996). Residues from therapeutants used in aquaculture may pose a serious threat to nontarget invertebrate communities in the receiving environment because many chemical treatments used in aquaculture are designed to eradicate invertebrate pathogens. The key concern, however, is the promotion of multiple resistance to antibiotics among the microbial community.

Chemicals Chemical use in aquaculture, including inorganic fertilizer, has been reviewed in a number of studies (Beveridge et al. 1991; Bergheim and Åsgård 1996; Boyd and Massaut 1999). Organic and inorganic fertilizer employed in semi-intensive pond aquaculture may significantly increase nutrient levels in the culture water. The minimal exchange of water in these systems limits the release of nutrients to the receiving environment (Edwards 1993). Lime is used extensively to condition pond sediments between production cycles, and impacts on the external environment are likely to be limited. More problematic may be the use of highly toxic and persistent chemicals, (e.g., chlorinated hydrocarbons and organotins) to eradicate predators, competitors, and disease vectors. Piscicides and molluscicides derived from plant extracts (e.g., rotenone, saponin, and nicotine) are widely used in tropical aquaculture (Baird 1994). However, Phillips et al. (1993) suggest that the environmental impact of using these products is expected to be limited because they readily biodegrade. The relatively nonspecific nature of these compounds and their possible impact on the health of workers during preparation and application are, however, of concern (Baird 1994).

Disinfectants (e.g., sodium hypochlorite, formalin, and benzalkonium chloride) used in production facilities, particularly hatcheries, may be discharged in aquaculture wastewater, but the consequences of this practice have not been investigated. Environmental impacts associated with the use of algaecides (e.g., copper sulfate) in shrimp ponds also have not been studied. Nevertheless, impacts depend on the application rate and the degree to which water is exchanged between the culture facility and receiving environment.

Aquaculture and Human Health Issues

As mentioned previously, products from small-scale aquaculture operations in nonindustrialized countries play an important role in the food security of poor and rural households. Small fish that may be cooked and eaten whole represent a more important source of several vitamins and minerals than equivalent portions taken from large fish. Therefore, even limited quantities of small fish, both stocked and small indigenous species, harvested from aquaculture systems can play an important role in the nutrition of poor households.

Operators The reuse of animal and human excreta and domestic and industrial wastewater for aquaculture is an important component in the management strategies adopted by vast numbers of poor households with limited access to alternative production-enhancing inputs. However, contact by operators with concentrated waste resources and continuous exposure to culture waters containing dilute waste concentrations may elevate risks associated with contracting communicable diseases associated with excreta and wastewater reuse. Therefore, risks posed by this activity must be considered. It should be noted that, compared with widespread practices of culturing fish in fecally polluted surface waters, the production of fish in correctly engineered and managed formal wastewater aquaculture systems may represent a lower level of risk (Pal and Das Gupta 1992). Furthermore, the correct design of wastewater aquaculture systems should facilitate rapid attenuation of pathogen numbers, limiting the risk posed to operators (Mara 1997).

Aquatic environments in nonindustrialized countries established for aquaculture may provide suitable habitats for pathogen hosts and disease vectors, principal among which are vectors of malaria and schistosomiasis. Where operators are in frequent contact with the culture water or where operators and their families live in close proximity to the culture facility, for example, on pond embankments or floating pontoons over cages, the risk of contracting diseases present in the culture system may increase significantly. Operators of aquatic farming systems may be at a greater risk from zoonosis when domestic animals and livestock are present. The common exploitation of aquatic resources for aquaculture and livestock husbandry and the deliberate integration of livestock with aquaculture systems to enhance resource flows in the farming system represent two common scenarios in which aquaculture and livestock production coexist. Aquatic habitats managed for aquaculture may harbor the hosts or vectors of several pathogens that infect domestic animals and livestock. These pathogens may then be passed to humans through a number of transmission routes (WHO 1999).

Consumers and Society Consumers in nonindustrialized countries eating produce from aquaculture operations using manure and wastewater resources may be exposed to a number of pathogens (i.e., trematodes and helminths); however, the risks are difficult to quantify. The prevalence of the pathogen in the population, the level of contamination in the culture system, the mode and degree of exposure of the consumer, and the resistance of the individual to infection may all require consideration when attempting to define the level of risk. Ubiquitous organisms (e.g., *E. coli* and *Salmonella* spp.) also represent possible hazards, but the level of risk may again be difficult to determine. In the vast majority of cases, products from aquaculture that are processed, handled, and cooked correctly will be microbiologically safe. Products that are not prepared and stored in an appropriate manner may constitute a possible hazard. Similarly in wild harvested species, failing to clean fish and shrimp properly in clean water may allow *Salmonella* to colonize the final product, and storing products incorrectly, for example, holding shellfish on market stalls, may permit poisonous bacteria to proliferate (Hatha et al. 1998).

Residues from soil and water conditioners, feed additives, and chemotherapeutants represent a potential hazard to consumers of products from aquaculture. Biocides, such as organochlorines, are sometimes used to harvest fish from ponds and small water bodies. It is possible that residues from these substances persist in fish eaten by consumers (Baird 1994).

It may be difficult to safeguard against exogenous contaminants entering the production system, especially where ecologically open production systems (e.g., flow-through ponds, cages, and pens) are being employed. However, where possible it is in the interests of the producer to safeguard the quality of products, not only to ensure that consumers continue to perceive the products as safe, but because products, especially from small-scale systems, are often consumed by the producers and their families. Ensuring that particularly perishable products from aquaculture retain their quality throughout the market chain and during processing represents a considerable challenge and one that is only likely to be met through the development of appropriate institutional structures and processes.

CONCLUSION

In western Europe and North America, social status, religious restrictions and the emergence of leisure classes during the industrial revolution have all shaped present-day inland aquaculture, whereas fish farming in inland areas of nonindustrialized countries has been overwhelmingly driven by the need to produce food (Beveridge and Little 2001). Aquaculture has emerged as an important source of food security in Asia. The types of species farmed also differ, for unlike temperate fresh waters, those in the tropics abound with fish that feed low in the food web (phytoplanktivores, zooplanktivores, and detrivores, e.g., carps, catfish, and tilapia) that can be readily integrated with crop and livestock production. Until very recently, freshwater fish farming in nonindustrialized countries was an integral part of the prevailing farming systems, unlike in North America and western Europe. However, intensification, in terms of higher stocking densities and a greater reliance on formulated, high-protein food, is evident in China and elsewhere. Mariculture practices in industrialized and nonindustrialized areas of the world have many features in common. Seaweeds and mollusks are universally raised by extensive methods and, with the exception of the fish species used in traditional coastal fish farming such as mullets or milkfish, farmed marine fish species are high-market-value carnivores grown in cages.

In conclusion, the role of the fish farmer is changing from merely raising fish to being an indispensable part of a chain for the production and delivery of safe, high-quality products to the consumer. There is little doubt that aquaculture production will become an increasingly important means of supplying aquatic products for human consumption. Food safety assurance measures and good aquacultural practices will need to be integrated into fish farm management programs. These should form an integral part of the fish "farm-to-table" food safety continuum.

REFERENCES

Abad, F.X., R.M. Pintó, R. Gajardo, and A. Bosch. 1997. Viruses in mussels: public health implications and depuration. J. Food Protection 60: 677–681.

Abbott, J.D., S.W. Hinton, and D.L. Borton. 1995. Pilot scale validation of the river/fish bioaccumulation modeling program for nonpolar hydrophobic organic compounds using the model compounds 2,3,7,8-TCDD and 2,3,7,8-TCDF. Environ. Toxicol. Chem. 14: 1999–2012.

Ackefors, H., V. Hilge, and O. Lindén. 1990. Contaminants in fish and shellfish products. *In*: Aquaculture Europe '89—Business Joins Science. Reviews and panel reports of the international conference, Bordeaux, France, October 2–4, 1989. Special Publication No. 12 (eds. N. De Pauw and R. Billard), pp. 305–344. Bredene, Belgium: European Aquaculture Society.

Agarwal, R.A. and D.K. Singh. 1988. Harmful gastropods and their control. Acta Hydrochim. Hydrobiol. 16: 113–138.

Ahmed, F.E. (ed.) (1991). *Seafood Safety*. Committee on Evaluation of the Safety of Fishery Products, Food and Nutrition Board, Institute of Medicine. Washington, DC: National Academy Press.

Amodio-Cocchieri, R., T. Cirillo, M. Amorena, M. Cavaliere, A. Lucisano, and U. Del Perte. 2000. Alkyltins in farmed fish and shellfish. Intl. J. Food Sci. Technol. 51: 47–151.

Arthington, A.H. and D.R. Bluhdorn. 1996. The effect of species introductions resulting from aquaculture operations. *In*: Aquaculture and water resource management (eds. D.J. Baird, M.C.M. Beveridge, L.A. Kelly, and J.F. Muir), pp. 114–139. Oxford, UK: Blackwell Science.

Bailey-Watts, A.E. 1994. Eutrophication. *In*: The Fresh Waters of Scotland (eds. P.S. Maitland, P.J. Boon, and D.S. McLusky), pp. 385–411. Chichester, UK: John Wiley and Sons.

Baird, D.J. 1994. Pest control in tropical aquaculture: an ecological hazard assessment of natural and synthetic control agents. Mitteilungen der Internationalen Vereinigung für Theoretische und Angewandte Limnologie. 24: 285–292.

Balasubramanian, S., M.R. Rajan, and S.P. Raj. 1992. Microbiology of fish grown in sewage-fed pond. Bioresource Technol. 40: 63–66.

Barber, M.C., L.A. Suárez, and R.R. Lassiter. 1991. Modelling bioaccumulation of organic pollutants in fish with an application to PCBs in Lake Ontario salmonids. Can. J. Fisheries Aquatic Sci. 48: 318–337.

Bardach, J.E. 1997. Aquaculture, pollution and biodiversity. *In*: Sustainable Aquaculture (ed. J.E. Bardach), pp. 87–99. Chichester, UK: John Wiley and Sons.

Bean, N.H. and M. Griffin. 1990. Foodborne disease outbreaks in the United States, 1973–1987: pathogens, vehicles, and trends. J. Food Protection 53: 804–817

Beardmore, J.A., G.C. Mair, and R.I. Lewis. 1997. Biodiversity in aquatic systems in relation to aquaculture. Aquaculture Res. 28: 829–839.

Berg, H., P. Michelsen, M. Troell, C. Folke, and N. Kautsky. 1996. Managing aquaculture for sustainability in tropical Lake Kariba, Zimbabwe. Ecological Economics 18: 141–159.

Bergheim, A. and T. Åsgård. 1996. Waste production from aquaculture. *In*: Aquaculture and Water Resource Management (eds. D.J. Baird, M.C.M. Beveridge, L.A. Kelly, and J.F. Muir), pp. 50–80. Oxford, UK: Blackwell Science.

Beril, C., J.M. Crance, F. Leguyader, V. Apaire-Marchais, F. Leveque, M. Albert, M.A. Goraguer, L. Schwartbrod, and S. Billadeul. 1996. Study of viral and bacterial indicators in cockles and mussels. Marine Pollution Bull. 32: 404–409

Beveridge, M.C.M. and D.C. Little. 2001. Aquaculture in traditional societies. *In*: Ecological Aquaculture (ed. B.A. Costa-Pierce). Blackwell, Oxford (in press).

Beveridge, M.C.M. and B.J. McAndrew. (eds.) 2000. *Tilapias: Biology and Exploitation.* p. 505. Dordrecht, The Netherlands: Kluwer.

Beveridge, M.C.M. and G. Haylor. 1998. Warm-water farmed species. *In*: Fish Farm Biology (eds. K. Black and A. Pickering), pp. 389–412. Sheffield, UK: Sheffield Academic Press.

Beveridge, M.C.M. and J.A. Stewart. 1998. Cage culture: limitations in lakes and reservoirs. *In*: Inland Fishery Enhancements (ed. T. Petr), FAO Fisheries Technical Paper, No. 374, pp. 263–279. Rome: Food and Agriculture Organization of the United Nations.

Beveridge, M.C.M., M.J. Phillips, and D.C. Macintosh. 1997. Aquaculture and the environment: the supply and demand for environmental goods and services by Asian aquaculture and the implications for sustainability. Aquaculture Res. 28: 101–111.

Beveridge, M.C.M. 1996. *Cage Aquaculture*. 2nd ed. Oxford, UK: Fishing News Books.

Beveridge, M.C.M. and M.J. Phillips. 1993. Environmental impact of tropical inland aquaculture. *In*: Environment and Aquaculture in Developing Countries (eds. R.S.V. Pullin, H. Rosenthal, and J.L. Maclean), pp. 213–236. Metro Manila, Philippines: International Center for Living Aquatic Resource Management.

Beveridge, M.C.M., M.J. Phillips, and R.M. Clarke. 1991. A quantitative and qualitative assessment of wastes from aquatic animal production. *In*: Aquaculture and Water Quality. Advances in World Aquaculture, Vol. 3 (eds. D.E. Brune and J.R. Tomasso), pp. 506–533. Baton Rouge, LA: The World Aquaculture Society.

Beveridge, M.C.M. 1984. The Environmental Impact of Freshwater Cage and Pen Fish Farming and the Use of Simple Models to Predict Carrying Capacity. FAO Technical Paper. No. 255. Rome: Food and Agriculture Organization of the United Nations.

Bhaskar, N., T.M.R. Setty, S. Mondal, M.A. Joseph, C.V. Rajul, B.S. Raghunath, and C.S. Anantha. 1998. Prevalence of bacteria of public health significance in cultured shrimp (*Penaeus monodon*). Food Microbiol. 15: 511–519.

Bhaskar, N., T.M.R. Setty, G.V.S. Reddy, Y.B. Manoj, C.S. Anantha, B.S. Raghunath, and J.M. Antony. 1995. Incidence of *Salmonella* in cultured shrimp *Penaeus monodon*. Aquaculture 138: 257–266.

Bhowmik, M.L., B.K. Pandey, and U.K. Sarkar. 1994. Microflora present in wastewater aquaculture ponds and fishes. Environ. Ecol. 12: 419–423.

Biddinger, G.R. and S.P. Gloss. 1984. The importance of trophic transfer in the bioaccumulation of chemical contaminants in aquatic ecosystems. Residue Revs. 91: 103–145.

Bloom, N.S. 1992. On the chemical form of mercury in edible fish and marine invertebrate tissue. Can. J. Fisheries Aquatic Sci. 49: 1010–1017.

Blum, D. and R.G. Feachem. 1985. *Health aspects of nightsoil and sludge use in agriculture and aquaculture. Part III. An epidemiological Perspective.* Dübendorf, Switzerland: International Reference Centre for Waste Disposal.

Borgmann, U. and D.M. Whittle. 1992. Bioenergetics and PCB, DDE, and mercury dynamics in Lake Ontario lake trout (*Salvelinus namaycush*): a model based on surveillance data. Can. J. Fisheries Aquatic Sci. 49: 1086–1096.

Boyd, C.E. and L. Massaut. 1999. Risks associated with the use of chemicals in pond aquaculture. Aquacultural Eng. 20: 113–132.

Brier, J.W. 1992. Emerging problems in seafood-borne parasitic zoonoses. Food Control 3: 2–7.

Bryan, F.L., J.J. Guzewich, and E.C.D. Todd. 1997. Surveillance of foodborne disease. III. Summary and presentation of data on vehicles and contributory factors; their value and limitations. J. Food Protection 60: 701–714.

Bryan, F.L. 1980. Epidemiology of foodborne diseases transmitted by fish, shellfish and marine crustaceans in the United States, 1970–1978. J. Food Protection 43: 859–876.

Bryan, F.L. 1977. Diseases transmitted by foods contaminated by wastewater. J. Food Protection 40: 45–56.

Buras, N. 1993. Microbial safety of produce from wastewater-fed aquaculture. *In*: Environment and Aquaculture in Developing Countries (eds. R.S.V. Pullin, H. Rosenthal, and J.L. Maclean), pp. 285–295. Metro Manila, Philippines: International Center for Living Aquatic Resource Management.

Buras, N., L. Duek, S. Niv, B. Hepher, and E. Sandbank. 1987. Microbiological aspects of fish grown in treated wastewater. Water Res. 21: 1–10.

Buras, N., L. Duek, and S. Niv. 1985. Reactions of fish to microorganisms in wastewater. Appl. Environ. Microbiol. 50: 989–995.

Chen, E.R. 1991. Current status of food-borne parasitic zoonoses in Taiwan. *In*: Emerging Problems in Food-Borne Parasitic Zoonosis: Impact on Agriculture and Public Health. Proceedings of the 33rd SEAMEO-TROPMED Regional Seminar, Chiang Mai, Thailand, 14–17 November 1990 (ed. J.H. Cross), pp. 62–64. Bangkok: SEAMEO Regional Tropical Medicine and Public Health Project, Thailand.

Coche, A.G. 1982. Cage culture of tilapias. *In*: Biology and Culture of Tilapias (eds. R.S.V. Pullin and R.H. Lowe-McConnell), pp. 205–246. Metro Manila, Philippines: International Center for Living Aquatic Resource Management.

Cole, J.G., D. Mackay, K.C. Jones, and R.E. Alcock. 1999. Interpreting, correlating, and predicting the multimedia concentration of PCDD/Fs in the United Kingdom. Environ. Sci. Technol. 33: 399–405.

Cooke, E.M. 1990. Epidemiology of foodborne illness: UK. Lancet 336: 790–793.

Costa-Pierce, B.A. 1996. Environmental impacts of nutrients from aquaculture: towards the evolution of sustainable aquaculture systems. *In*: Aquaculture and Water Resource Management (eds. D.J. Baird, M.C.M. Beveridge, L.A. Kelly, and J.F. Muir), pp. 81–113. Oxford, UK: Blackwell Science.

Cross, J.H. 1991. Emerging Problems in Food-Borne Parasitic Zoonosis: Impact on Agriculture and Public Health. Proceedings of the 33rd SEAMEO-TROPMED Regional Seminar, Chiang Mai, Thailand, 14–17 November 1990. Bangkok: SEAMEO Regional Tropical Medicine and Public Health Project, Thailand.

Cross, J.H. 1984. Changing patterns of some trematode infections in Asia. Arzneimittel forschung. 34(II), Nr. 9b: 1224–1226.

Cross, P. 1985. Health aspects of nightsoil and sludge use in agriculture and aquaculture. Part I. Existing practices and beliefs in the utilization of human excreta. Report No. 04/85. Dübendorf, Switzerland: International Reference Centre for Waste Disposal.

Dallinger, R., F. Prosi, H. Segner, and H. Back. 1987. Contaminated food and uptake of heavy metals by fish: a review and a proposal for further research. Oecologia 73: 91–98.

Dalsgaard, A. 1998. The occurrence of human pathogenic *Vibrio* spp. and *Salmonella* in aquaculture. Intl. J. Food Sci. Technol. 33: 127–138.

Dalsgaard, A. and J.E. Olsen. 1995. Prevalence of *Salmonella* in dry pelleted chicken manure samples obtained from shrimp farms in a major shrimp production area in Thailand. Aquaculture 136: 291–295.

Davies, I.M. and J.C. McKie. 1987. Accumulation of total tin and tributyltin in muscle tissue of farmed Atlantic salmon. Marine Pollution Bull. 18: 405–407.

Deardorff, T.L. and R.M. Overstreet. 1991. Seafood-transmitted zoonoses in the United States: the fishes, the dishes, and the worms. *In*: Microbiology of Marine Food Products (eds. D.R. Ward and C. Hackney), pp. 211–265. New York: Van Nostrand Reinhold.

De Boer, J. and H. Pieters. 1991. Dietary accumulation of polychlorinated biphenyls, chlorinated pesticides and mercury by cultivated eels, *Anguilla anguilla*. L. Aquaculture Fisheries Management 22: 329–334.

De Grave, S., S.J. Moore, and G. Burnell. 1998. Changes in benthic macrofauna associated with intertidal oyster, *Crassostrea gigas* (Thunberg) culture. J. Shellfish Res. 17: 1137–1142.

Delzell, E., J. Giesy, J. Doull, D. Mackay, I. Munro, and G. Williams, eds. 1994. Interpretive review of the potential adverse effects of chlorinated organic chemicals on human health and the environment. Report of an Expert Panel. Regulatory Toxicology and Pharmacology. 20(1), part 2.

Ditrich, O., T. Scholtz, and M. Giboda. 1990. Occurrence of some medically important flukes (Trematoda: Opisthorchiidae and Heterophyidae) in Nam Ngum water reservoir, Laos. Southeast Asian J. Trop. Med. Pub. Hlth. 21: 482–488.

Dombroski, C.S., L.-A. Jaykus, and D.P. Green. 1999. Occurrence and control of *Vibrio vulnificus* in shellfish. J. Aquatic Food Product Technol. 8: 11–25.

Edwards, P. 1990. General discussion on wastewater-fed aquaculture. *In*: Wastewater-Fed Aquaculture. Proceedings of the International Seminar on Wastewater Reclamation and Reuse for Aquaculture, Calcutta, India, 6–9 December 1988 (eds. P. Edwards and R.S.V. Pullin), pp. 281–291. Bangkok: Environmental Information Center, Asian Institute of Technology.

Edwards, P. 1993. Environmental issues in integrated agriculture-aquaculture and wastewater-fed fish culture systems. *In*: Environment and Aquaculture in Developing Countries (eds. R.S.V. Pullin, H. Rosenthal, and J.L. Maclean), pp. 139–170. Metro Manila, Philippines: International Center for Living Aquatic Resource Management.

Edwards, P. 1980. A review of recycling organic wastes into fish, with emphasis on the tropics. Aquaculture 21: 261–279.

Edwards, P. and R.S.V. Pullin (eds). 1990. *Wastewater-Fed Aquaculture*. Proceedings of the International Seminar on Wastewater Reclamation and Reuse for Aquaculture, Calcutta, India, 6–9 December 1988. Bangkok: Environmental Information Center, Asian Institute of Technology.

Eves, A., C. Turner, A. Yakupitayage, N. Tongdee, and S. Ponza. 1995. The microbiological and sensory quality of septage-raised Nile tilapia (*Oreochromis niloticus*). Aquaculture 132: 261–272.

FAO. 2000. Fishstat+, version 2.3. Downloadable database from www.fao.org/fi/statist/fisoft/fishplus.asp. Rome: Food and Agriculture Organization of the United Nations.

Fernàndez, H.R. and L.A. Fernàndez. 1998. Introduction of rainbow trout in Tucuman Province, Argentina: problems and solutions. Ambio 27: 584–585.

Fisher, D.J., J.R. Clark, M.H. Roberts, J.P. Connolly, and L.H. Mueller. 1986. Bioaccumulation of kepone by spot (*Leiostomus xanthurus*): importance of dietary accumulation and ingestion rate. Aquatic Toxicity 9: 161–178.

Flaherty, M., B. Szuster, and P. Miller. 2000. Low salinity inland shrimp farming in Thailand. Ambio 29: 174–179.

Furedy, C. 1990. Social aspects of human excreta reuse: implications for aquacultural projects in Asia. *In*: Wastewater-Fed Aquaculture. Proceedings of the International Seminar on Wastewater Reclamation and Reuse for Aquaculture, Calcutta, India, 6–9 December 1988 (eds. P. Edwards and R.S.V. Pullin), pp. 251–266. Bangkok: Environmental Information Center, Asian Institute of Technology.

Gerba, C.P. 1988. Viral disease transmission by seafoods. Food Technol. 42(3): 99–103.

GESAMP. 1987. *IMO/FAO/UNESCO/WMO/IAEA/UN/UNEP Joint Group of Experts on the Scientific Aspects of Marine Pollution. Land/Sea Boundary Flux of Contaminants: Contribution from Rivers. Reports and Studies GESAMP 32.* London: International Maritime Organization.

GESAMP. 1986. *IMO/FAO/UNESCO/WMO/IAEA/UN/UNEP Joint Group of Experts on the Scientific Aspects of Marine Pollution. Review of Poten-*

tially Harmful Substances. Arsenic, Mercury and Selenium. Reports and Studies GESAMP 28. London: International Maritime Organization.

GESAMP. 1985. *IMO/FAO/UNESCO/WMO/IAEA/UN/UNEP Joint Group of Experts on the Scientific Aspects of Marine Pollution. Review of Potentially Harmful Substances. Cadmium, Lead and Zinc. Reports and Studies GESAMP 22.* London: International Maritime Organization.

Gibson, D.M. 1992. Pathogenic microorganisms of importance in seafoods. *In*: Quality Assurance in the Fish Industry, (eds. H.H. Huss, M. Jakobsen, and J. Liston), pp. 197–209. Amsterdam: Elsevier Science Publishers.

Gobas, F.A.P.C. and J.A. McCorquodale (eds). 1992. *Chemical dynamics in fresh water ecosystems*. Chelsea, MI: Lewis Publishers.

Gowen, R.J., 1994. Managing eutrophication associated with aquaculture development. J. Appl. Ichthyology 10: 242–257.

Guerrin, F., V. Burgat-Sacaze, and P. de Saqui-Sannes. 1990. Levels of heavy metals and organochlorine pesticides of cyprinid fish reared four years in a wastewater treatment pond. Bull. Environ. Contamination Toxicol. 44: 461–467.

Guzewich, J.J., F.L. Bryan, and E.C.D. Todd. 1997. Surveillance of foodborne disease. I. Purposes and types of surveillance systems and networks. J. Food Protection 60: 555–566.

Guzewich, J.J. and D.L. Morse. 1986. Sources of shellfish in outbreaks of probable viral gastroenteritis: implications for control. J. Food Protection 49: 389–394.

Håkonson, L. 1980. The quantitative impact of pH, bioproduction and Hg contamination on the Hg content of fish (pike). Environ. Pollution (Series B). 1: 285–304.

Hall, L.W. and A.E. Pinkey. 1985. Acute and sublethal effects of organotin compounds on aquatic biota: an interpretive literature evaluation. Crit. Rev. Toxicol. 14: 159–209.

Hansen, P.-D. 1980. Uptake and transfer of the chlorinated hydrocarbon lindane (γ-BHC) in a laboratory freshwater food chain. Environ. Pollution (series A). 21: 97–108.

Hatha, A.A.M., N. Paul, and B. Rao. 1998. Bacteriological quality of individually quick-frozen (IQF) raw and cooked ready-to-eat shrimp produced from farm raised black tiger shrimp (*Penaeus monodon*). Food Microbiol. 15: 177–183.

Hilton, J.W., P.V. Hodson, H.E. Braun, J.L. Leatherland, and S.J. Slinger. 1983. Contaminant accumulation and physiological response in rainbow trout (*Salmo gairdneri*) reared on naturally contaminated diets. Can. J. Fisheries Aquatic Sci. 40: 1987–1994.

Hirazawa, N., T. Ohtaka, and H. Hata. 1988. Challenge trials on the anthelmintic effect of drugs and natural agents against the monogenean *Heterobothrium okamotoi* in the tiger puffer *Takifugu rubripes*. Aquaculture 188: 1–13.

Howgate, P. 1998. Review of public health safety of products from aquaculture. Intl. J. Food Sci. Technol. 33: 99–125.

Huss, H.H., P. Dalgaard, and L. Gram. 1997. Microbiology of fish and fish products. *In*: Seafood from Producer to Consumer: Integrated Approach to Quality (eds. J.B. Luten, T. Børrensen, and J. Oehlenschläger), pp. 413–430. Amsterdam: Elsevier.

Huss, H.H. 1994. Assurance of seafood quality. FAO Fisheries Technical Paper. No. 334. Rome: Food and Agriculture Organization of the United Nations.

Inglis, V., R.H. Richards, and K.N. Woodward. 1993. Public health aspects of bacterial infections of fish. *In*: Bacterial Diseases of Fish (ed. N.R. Bromage), pp. 284–303. Oxford, UK: Blackwell.

Institute for Tropical Endemic Diseases. 1984. Proceedings of the International Symposium on Human Trematode Infections in Southeast and East Asia, Kyongju, Republic of Korea, October 19–21, 1983. Arzneimittelforschung 34, Nr. 9b.

IRCWD. 1990. Human waste use in agriculture and aquaculture. Utilization practices and health perspectives. IRCWD Report No 09/90. Dübendorf, Switzerland: International Reference Centre for Waste Disposal.

IRCWD. 1988. Health aspects of wastewater and excreta use in agriculture and aquaculture. IRCWD News, 24/25. Dübendorf, Switzerland: International Reference Centre for Waste Disposal.

IRCWD. 1985. Health aspects of wastewater and excreta use in agriculture and aquaculture. IRCWD News, 23. Dübendorf, Switzerland: International Reference Centre for Waste Disposal.

Jackson, L.J. and D.E. Schindler. 1996. Field estimates of net trophic transfer of PCBs from prey fishes to lake Michigan salmonids. Environ. Sci. Technol. 30: 1861–1865.

Jones, J.G. 1990. Pollution from fish farms. J. Inst. Water Environ. Mgmt. 4: 14–18.

Joseph, M.M. 1998. Mussel and oyster culture in the tropics. *In*: Tropical Mariculture (ed. S.S. de Silva), pp. 309–360. London: Academic Press.

Karakassis, I., E. Hatziyanni, M. Tsapakis, and W. Plaiti. 1999. Benthic recovery following cessation of fish farming: a series of successes and catastrophes. Marine Ecology Prog. Ser. 184: 205–218.

Kino, H., H. Inaba, H.V. De, L.V. Chau, D.T. Son, H.T. Hao, N.D. Toan, L.D. Cong, and M. Sano. 1998. Epidemiology of Clonorchiasis in Ninh Binh province, Vietnam. Southeast Asian J. Trop. Med. Publ. Hlth. 29: 250–254.

Lehmann, I. 1993. *Contents of Environmental Contaminants in Fish from the New States of Germany*. Arbeiten aus dem Institut für Biochemie und Technologie der Bundesforschungsanstalt für Fischerei, Nr. 7, Hamburg: Germany.

Lieb, A.J., D.D. Bills, and R.O. Sinnhuber. 1974. Accumulation of dietary polychlorinated biphenyls (Aroclor 1254) by rainbow trout (*Salmo gairdneri*). J. Agric. Food Chem. 22: 638–642.

Lima dos Santos, C.A.M. 1997. The possible use of HACCP in the prevention and control of food-borne trematode infections in aquacultured fish. *In*: Seafood Safety, Processing, and Biotechnology (eds. F. Shahidi, Y. Jones, and D.D. Kitts), pp. 53–64. Lancaster, PA: Technomic Publishing Inc.

Mackay. 1991. Multimedia Environmental Models: The Fugacity Approach. Chelsea, MI: Lewis Publishers.

Makarewicz, J.C., J.K. Buttner, and T.W. Lewis. 1993. Uptake and retention of mirex by fish maintained on formulated and natural diets in Lake Ontario waters. Progressive Fish-Culturist. 55: 163–168.

Malek, E.A. 1980. *Snail-Transmitted Parasitic Diseases, vol II*. Boca Raton, FL: CRC Press.

Mara, D. 1997. *Design Manual for Waste Stabilization Ponds in India*. Leeds, UK: Lagoon Technology International.

McCullough, F.S. 1990. Schistosomiasis and aquaculture. *In*: Wastewater-Fed Aquaculture. Proceedings of the International Seminar on Wastewater Reclamation and Reuse for Aquaculture, Calcutta, India, 6–9 December 1988 (eds. P. Edwards and R.S.V. Pullin), pp. 237–249. Bangkok: Environmental Information Center, Asian Institute of Technology.

Mitchell, A.J. 1995. Importance of treatment duration for praziquantel used against larval digenetic trematodes in sunshine bass. J. Aquatic Anim. Hlth. 7: 327–330.

Naegel, L.C.A. 1990. A review of public health problems associated with the integration of animal husbandry and aquaculture, with emphasis on southeast Asia. Biological Wastes 31: 69–83.

Nedoluha, P.C. and D. Westhoff. 1993. Microbiological flora of aquacultured hybrid striped bass. J. Food Protection 56: 1054–1060.

New, M.B. and S. Singholka. 1985. Freshwater Prawn Farming. FAO Fisheries Technical Paper. No. 255. Rome: Food and Agriculture Organization of the United Nations.

Nickerson, D.J. 1999. Trade-off of mangrove area development in the Philippines. Ecol. Econ. 28: 279–298.

Ninawe, A.S. 1994. Effective waste management. Fish Farmer, July/August, 1994: 34–37.

Notermans, S. and M. Borgdorff. 1997. A global perspective of foodborne disease. J. Food Protection 60: 1395–1399.

Ogbondeminu, F.S. 1993a. The occurrence and distribution of enteric bacteria in fish and water of tropical aquaculture ponds in Nigeria. J. Aquaculture Tropics 8: 61–66.

Ogbondeminu, F.S. 1993b. Health significance of gram-negative bacteria associated with waste-fed tropical aquaculture system. Intl. J. Environ. Hlth. Res. 3: 10–17.

Oláh, J., N. Sharangi, and N.C. Datta. 1986. City sewage fish ponds in Hungary and India. Aquaculture 54: 129–134.

Páez-Osuna, F., S.R. Guerrero-Galván, and A.C. Ruiz-Fernández. 1998. The environmental impact of shrimp aquaculture and the coastal pollution in Mexico. Marine Pollution Bull. 36: 65–75.

Pal, D. and C. Das Gupta. 1992. Microbial pollution in water and its effect on fish. J. Aquatic Anim. Hlth. 4: 32–39.

Pandey, B.K., U.K. Sarkar, M.L. Bhowmik, and S.D. Tripathi. 1995. Accumulation of heavy metals in soil, water, aquatic weed and fish samples of sewage-fed ponds. J. Environ. Biol. 16: 97–103.

Park, M.-S., S.-W. Kim, Y.-S. Yang, C.-H. Park, W.T. Lee, C.U. Kim, E.-M. Lee, S.-U. Lee, and S. Huh. 1993. Intestinal parasite infections in the inhabitants along the Hantan River, Chorwon-gu. Korean J. Parasitol. 31: 375–378.

Payne, A.I. 1984. Use of sewage waste in warm water aquaculture. *In*: Reuse of Sewage Effluent. pp. 157–171. London: Thomas Telford.

Perry, A.S., A. Gasith, and Y. Mozel, 1983. Pesticide residues in fish and aquatic invertebrates. Arch. Toxicol. Suppl. 6: 199–204.

Phillips, D.J.H. 1995. The chemistries and environmental fates of trace metals and organochlorines in aquatic ecosystems. Marine Pollution Bull. 31: 193–200.

Phillips, D.J.H. 1993. Developing-country aquaculture, trace chemical contaminants, and public health concerns. *In*: Environment and Aquaculture in Developing Countries (eds. R.S.V. Pullin, H. Rosenthal, and J.L. Maclean), pp. 296–311. Metro Manila: International Center for Living Aquatic Resource Management.

Phillips, M.J. 1998. Tropical mariculture and coastal environmental integrity. *In*: Tropical Mariculture (ed. S.S. de Silva), pp. 17–70. London: Academic Press.

Phillips, M.J., C. Kwei Lin, and M.C.M. Beveridge. 1993. Shrimp culture and the environment: lessons from the world's most rapidly expanding warmwater aquaculture sector. *In*: Environment and Aquaculture in Developing Countries (eds. R.S.V. Pullin, H. Rosenthal, and J.L. Maclean), pp. 171–197. Metro Manila: International Center for Living Aquatic Resource Management.

Pillay, T.V.R. 1992. *Aquaculture and the Environment.* Farnham, UK: Fishing News (Books).

Prein, M. 1990. Wastewater-fed fish culture in Germany. *In*: Wastewater-Fed Aquaculture. Proceedings of the International Seminar on Wastewater Reclamation and Reuse for Aquaculture, Calcutta, India, 6–9 December 1988 (eds. P. Edwards and R.S.V. Pullin), pp. 13–47. Bangkok: Environmental Information Center, Asian Institute of Technology.

Primavera, J.H. 1998. Tropical shrimp farming and its sustainability. *In*: Tropical Mariculture (ed. S.S. de Silva), pp. 257–289. London: Academic Press.

Primavera, J.H. 1991. Intensive prawn farming in the Philippines: ecological, social, and economic implications. Ambio 20: 28–33.

Pullin, R.S.V. and M. Prein. 1995. Fishponds facilitate natural resources management on small-scale farms in tropical developing countries. *In*: Proceedings of the seminar on the management of integrated freshwater agro-piscicultural ecosystems in tropical areas (eds. J.J. Symoens and J.C. Micha), pp. 169–186. Brussels, Belgium: Technical Centre for Agricultural and Rural Cooperation and Royal Academy of Overseas Sciences.

Pullin, R.S.V., H. Rosenthal, and J.L. Maclean (eds). 1993. *Environment and Aquaculture in Developing Countries*. Metro Manila: International Center for Living Aquatic Resource Management. p. 295.

Rajan, M.R., S. Balasubramanian, and S.P. Raj. 1995. Accumulation of heavy metals in sewage-grown fishes. Bioresource Technol. 52: 41–43.

Reilly, A. and F. Käferstein. 1997. Food safety hazards and the application of the principles of the hazard analysis and critical control point (HACCP) system for their control in aquaculture production. Aquaculture Res. 28: 735–752.

Reilly, P.J.A. and D.R. Twiddy. 1992. *Salmonella* and *Vibrio cholerae* in brackish water cultured tropical prawns. Intl. J. Food Microbiol. 16: 293–301.

Reilly, P.J.A., D.R. Twiddy, and R.S. Fuchs. 1992. Review of the occurrence of Salmonella in cultured tropical shrimp. FAO Fisheries Circular No. 851. Rome: Food and Agriculture Organization of the United Nations.

Rim, H.-J. 1982. Clonorchiasis. *In*: *CRC Handbook Series in Zoonoses, Section C: Parasitic Zoonoses* (eds. G.V. Hillyer and C.F. Hopla), pp. 17–32. Boca Raton, FL: CRC Press.

Saheki, K., S. Kobayashi, and T. Kawanishi. 1989. Salmonella contamination of eel culture ponds. Nippon Suisan Gakkaishi 44: 675–679. (In Japanese).

Sakanar, J.A. and J.H. McKerrow. 1989. Anisakiasis. Clin. Microbiol. Rev. 2: 278–284.

Santiago, A. 1995. The ecological impact of tilapia cage culture in Sampaloc lake, Philippines. *Proceedings of the Third Asian Fisheries Symposium*. pp. 170–186. Singapore: Asian Fisheries Society.

Schweinfurth, H.A. and P. Günzel. 1987. The tributyltins: Mammalian toxicity and risk evaluation for humans. *In*: Proceedings of the Oceans '87 Symposium, Vol 4, Organotin Symposium. pp. 1421–1437. Washington, DC: Marine Technical Society.

Semesi, A.K. 1998. Mangrove management and utilization in Eastern Africa. Ambio 27: 620–626.

Sevilleja, R.C., E.V. Manalili, and R.D. Guerrero (eds). 1993. *Reservoir Fisheries Management and Development in the Philippines*. Philippine Council for Aquatic and Marine Research and Development, Los Banos, Philippines. pp. 170–186. Brussels: CTA/Royal Academy of Overseas Sciences.

Sharma, B.K. and J. Oláh. 1986. Integrated fish-pig farming in India and Hungary. Aquaculture 54: 135–139.

Short, J.W. and F.P. Thrower. 1986. Accumulation of butyltins in muscle tissue of Chinook salmon reared in sea pens treated with tri-*n*-butyltin. Marine Pollution Bull. 12: 542–545.

Sin, A.W. and M.T.L. Chiu. 1987. The culture of silver carp, bighead, grass carp and common carp in secondary effluents of a pilot sewage treatment plant. Resources Conservation. 13: 231–246.

Slabbert, J.L., W.S.G. Morgan, and A. Wood. 1989. Microbiological aspects of fish cultured in wastewaters—the South African experience. Water Sci. Technol. 21: 307–310.

Slootweg, R., E.A. Malek, and F.S. McCullough. 1994. The biological control of snail intermediate hosts of schistosomiasis by fish. Rev. Fish Biol. Fisheries 4: 67–90.

Stenton-Dozey, J.M.E., L.F. Jackson, and A.J. Busby. 1999. Impact of mussel culture on macrobenthic community structure in Saldanha Bay, South Africa. Marine Pollution Bull. 39: 357–366.

Strauss, M. 1996. *Health (pathogen) considerations regarding the use of human waste in aquaculture*. Environmental Research Forum, Vol. 5–6, 83–98. Switzerland: Transtec Publications.

Strauss, M. 1985. *Health aspects of nightsoil and sludge use in agriculture and aquaculture. Part II. Pathogen survival.* Report No. 04/85. Dübendorf, Switzerland: International Reference Centre for Waste Disposal.

Tarifeño-Silva, E., L.Y. Kawasaki, D.P. Yu, M.S. Gordon, and D.J. Chapman. 1982. Aquacultural approaches to recycling of dissolved nutrients in secondarily treated domestic wastewaters. III. Uptake of dissolved heavy metals by food chains. Water Res. 16: 59–65.

Tatem, H.E. 1986. Bioaccumulation of polychlorinated biphenyls and metals from contaminated sediment by freshwater prawns, *Macrobrachium rosenbergii* and clams, *Corbicula fluminea.* Arch. Environ. Contamination Toxicol. 15: 171–183.

Todd, E. 1990. Epidemiology of foodborne illness: North America. Lancet 336: 788–790.

Todd, E.C.D. 1992. Foodborne disease in Canada—a 10 year summary from 1975–1984. J. Food Protection 55: 123–132.

Tsuda, T., S. Aoki, T. Inoue, and M. Kojima. 1994. Accumulation and excretion of pesticides used as insecticides or fungicides in agricultural products by the willow shiner *Gnathopogon caerulescens*. Comptes Biochem. Physiol. 107C: 469–473.

Turner, J.W.D., R.R. Sibbald, and J. Hemens. 1986. Chlorinated secondary domestic sewage effluent as a fertilizer for marine aquaculture. III. Assessment of bacterial and viral quality and accumulation of heavy metals and chlorinated pesticides in culture fish and prawns. Aquaculture 53: 157–168.

Van den Heever, D.J. and B.J. Frey. 1994. Microbiological quality of the catfish (*Clarius gariepinus*) kept in treated waste water and natural dam water. Water SA 20: 113–118.

Vaughn, J.M. and E.F. Landry. 1984. Public health considerations associated with molluscan aquaculture systems: human viruses. Aquaculture 39: 299–315.

Vicharsi, S., V. Viyanant, and E.S. Upatham. 1982. *Opisthorchis viverrini*: intensity and rates of infection in cyprinoid fish from an endemic focus in Northeast Thailand. Southeast Asian J. Trop. Med. Publ. Hlth. 13: 138–141.

Vojinovic-Miloradov, M., P. Marjanovic, D. Buzarov, S. Pavkov, Lj. Dimitrijevic, and M. Miloradov. 1992. Bioaccumulation of polychlorinated biphenyls and organochlorine pesticides in selected fish species as an indicator of pollution in Vojvodina, Yugoslavia. Water Sci. Technol. 26: 2361–2364.

Ward, D.R. and C. Hackney. 1991. *Microbiology of Marine Food Products*. New York: Van Nostrand Reinhold.

Welcomme, R.L. 1988. *International introductions of inland aquatic species*. FAO Fisheries Technical Paper. No. 294. Rome: Food and Agriculture Organization of the United Nations.

Wells, D.E. and J. de Boer. 1999. Polychlorinated biphenyls, dioxins and other polyhalogenated hydrocarbons as environmental contaminants in food. *In*: *Environmental Contaminants in Food* (eds. C.F. Moffat and K.J. Whittle), pp. 305–363. Sheffield, UK: Sheffield Academic Press.

Weston, D.P. 1996. Environmental considerations in the use of antibacterial drugs in aquaculture. *In*: Aquaculture and Water Resource Management (eds. D.J. Baird, M.C.M. Beveridge, L.A. Kelly, and J.F. Muir), pp. 140–165. Oxford, UK: Blackwell Science.

WHO. 1999. Food safety issues associated with products of aquaculture. Report of a joint FAO/NACA/WHO Study Group. WHO Technical Report Series 883. Geneva: World Health Organization.

WHO. 1995. Control of Foodborne Trematode Infections. WHO Technical Report Series 849. Geneva: World Health Organization.

WHO. 1989a. Health guidelines for the use of wastewater in agriculture and aquaculture. WHO Technical Report Series 778. Geneva: World Health Organization.

WHO. 1989b. Mercury—environmental aspects. Environmental Health Criteria 86. Geneva: World Health Organization.

Wiener, J.G. and J.P. Giesy. 1978. Concentrations of Cd, Cu, Mn, Pb, and Zn in fishes in highly organic softwater pond. J. Fisheries Res. Board Canada 36: 270–279.

Wong Chor Yee, A. 1999. New developments in integrated dike-pond agriculture-aquaculture in the Zhujiang Delta, China: ecological implications. Ambio. 28: 529–533.

Wyatt, L.E., R. Nickelson, and C. Vanderzant. 1979. *Edwardsiella tarda* in freshwater catfish and their environment. Appl. Environ. Microbiol. 38: 710–714.

Yu, J.-R., S.-O. Kwon, and S.-H. Lee. 1994a. Clonorchiasis and metagonimiasis in the inhabitants along the Talchongang River, Cungwan-gun. Korean J. Parasitol. 32: 267–269.

Yu, S.-H., L.-Q. Xu, Z.-X. Jiang, S.-H. Xu, J.-J. Han, Y.-G. Zhu, J. Chang, J.-X. Lin, and N. Xu. 1994b. Nationwide survey of human parasites in China. Southeast Asian J. Trop. Med. Publ. Hlth. 25: 4–10.

3

Public, Animal, and Environmental Aquaculture Health Issues in Industrialized Countries

Michael L. Jahncke and Michael H. Schwarz

INTRODUCTION

The aquaculture industry in industrialized countries is maturing and is under increased scrutiny and regulations. Thus it is important that aquaculture producers identify all potential risks associated with their operations. An integrated risk analysis approach consisting of risk assessment, management, and communication is essential to minimize unintended consequences from aquaculture. Such an integrated approach should build on the totality of issues and past aquaculture experiences so that it neither causes nor gives the appearance of contributing to unacceptable risks to the public, animals, and environment.

Public, Animal, and Environmental Aquaculture Health Issues,
Edited by Michael L. Jahncke, E. Spencer Garrett, Alan Reilly,
Roy E. Martin, and Emille Cole.
ISBN 0-471-38772-X (cloth)

Much of the world's commercial fisheries are either declining or are on the verge of overexploitation because of the world's expanding population putting increased pressure on the oceans' animal and plant resources (NMFS 2000a; FAO 2000). Responsible aquaculture operations can successfully provide safe and wholesome products to supplement the declining supply of wild-caught fishery products. The future of aquaculture is bright; and aquacultured products are as safe and wholesome as wild-caught species. Industrialized countries can no longer rely solely on domestically caught wild species to meet their seafood demands. The desire for high-quality fishery products will continue to increase over the foreseeable future, exceeding that which can be supplied by commercial harvests of wild species (Garrett et al. 2000; FAO 2000).

This chapter discusses public, animal, and environmental health issues from an industrialized country perspective. Public health issues include bacterial pathogens, viruses, parasites, the use of antibiotics in aquaculture, and the potential for development of antibiotic-resistant strains of bacteria. Animal and environmental health issues address bacteria, viruses, parasites, antibiotic residues, escapements, nutrient loading levels, siting issues, effluents, etc. Examples of significant diseases, parasites, and environmental issues are presented in this chapter along with possible control options that can be effective in managing these issues. The next to last section of this chapter presents two case studies that provide examples of how to establish successful aquaculture operations that address public, animal, and environmental health issues. The first case study is of a large commercial hybrid striped bass aquaculture firm located in the state of California, USA. The second case study is of a salmon net pen rearing operation in Ireland. Sustainable aquaculture must be based on a holistic approach that includes technology knowledge and best management practices and addresses the basic issues of environmental, animal, and public health (Fig. 3.1).

AN OVERVIEW OF AQUACULTURE SYSTEMS

Major aquaculture production in industrialized countries occurs in ponds, flow-through systems, net pens, bottom leases, and closed recirculating aquaculture systems (RAS). To date, most aquaculture production in industrialized countries has occurred in ponds, net pens, bottom leases, and flow-through raceways. However, because of limited land and water resources and multiple-user conflicts in coastal areas,

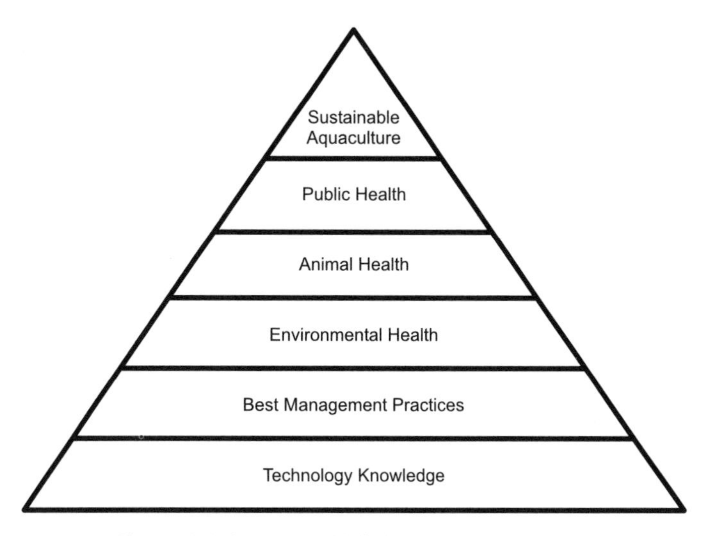

Figure 3.1. *Integrated Holistic Aquaculture Approach*

the trend in development of new aquaculture operations is offshore net pens and land-based RAS.

Aquaculture production ponds can be either man-made or natural bodies of water, usually less than 10 ha, used for the culture of aquatic organisms. Flow-through operations can be defined as raceways with a single pass of water. The raceways are designed to operate in series or parallel, dependent on species requirements and land and water resource availability, and are being used in industrialized countries to raise high-value marine species. Bottom leases are used to culture bivalve molluscan shellfish and are located near the shore in areas protected from high winds and wave action. RAS are aquaculture systems with at least partial water reuse and are analogous to land-based agriculture production systems such as swine barns or poultry broiler houses. RAS allows multiple reuse of water within a tank system, allowing high-density production in a limited area.

Catfish pond production and bottom bivalve culture dominate the US aquaculture industry, whereas Canada's aquaculture industry consists primarily of marine offshore net pens, nearshore bottom leases, and coastal area flow-through systems. Japan and industrialized European aquaculture production is primarily comprised of bottom culture techniques for bivalve molluscan shellfish and net pen production of finfish (FAO 1997). Likewise, Australia and New Zealand aquaculture industries concentrate on bottom culture for oysters and net pens for finfish (FAO 1997).

Advantages and Disadvantages of Aquaculture Systems

Ponds are the simplest and most traditional form of aquaculture production systems. The primary advantages are the lower production and labor costs associated with pond culture operations. Disadvantages of pond systems include higher land and water resource requirements, competing commercial uses for the land, fewer disease management options, loss of cultured species to birds and other animals, and limited control of possible chemical contamination of ponds from surrounding agricultural operations. Furthermore, pond wastewater effluents are discharged directly into the environment, posing potential disease risks to wild species and discharging nutrient effluents into the surrounding environment. Current research efforts are underway to develop aquaculture pond operations with minimal or zero water discharges.

Flow-through systems (e.g., tank/raceway production) are located in areas where abundant quantities of fresh water are available or along coastal areas for culture of marine species. Advantages of flow-through systems include reduced land requirements and relatively low production costs. Flow-through systems also allow for easy application of therapeutants for disease treatment of the cultured organisms.

Disadvantages of flow-through systems include reliance on large quantities of water, predation of cultured finfish by birds and other animals, possible escapements of cultured animals, and direct transfer of waste effluents into the environment. Flow-through production facilities are beginning to reuse water and develop remedial solids capture/removal systems.

Aquaculture production in net pens has historically been associated with culture of aquatic species in lakes, coastal bays, or protected nearshore environments. Because of environmental concerns and multiple-user resource conflicts, net pens are now being located further offshore to allow tidal or windblown currents to remove effluent wastes. Issues of concern include possible disease transmission from cultured to wild species, escapements of cultured species into the environment, and impacts of nutrients and waste effluents on benthic and other aquatic organisms.

Bottom lease systems are used primarily to culture bivalve molluscan species. The lower costs associated with these methods are an advantage. Disadvantages include little direct control of water quality and limited protection of cultured shellfish from predators and disease organisms.

RAS is an alternative production method used in locations that are not readily amenable to ponds or raceways. Aquaculture production in

RAS incorporates water pumping, solids removal, biofiltration, and oxygenation/degassing equipment to recycle culture water over and over. Advantages of this type include reduced reliance on land and water resources, maximized stock security and environmental control, and a concentrated waste stream that is amenable to conventional wastewater treatment facilities (Jahncke and Schwarz 1998). Public and animal health can be optimized in RAS by completely enclosing the system within a building to eliminate nonaquatic disease vectors and prefiltering/sterilizing/disinfecting all incoming water to eliminate waterborne pathogens and pollutants. The addition of therapeutants for disease control is also easier compared with other production systems. A major disadvantage of RAS is the higher production costs due to higher infrastructure and energy costs. To offset higher costs, the RAS industry is focusing on culturing higher-value biological species and maximizing system carrying capacities.

OVERVIEW OF AQUACULTURE PRODUCTION

Only 13% of the total worldwide aquaculture production occurred in industrialized countries during 1998 (FAO 2000). The following two tables rank the aquaculture production of 20 industrialized countries (Table 3.1) and the leading aquacultured species produced in these countries (Table 3.2) (FAO 2000).

PUBLIC HEALTH ISSUES

Properly applied, aquaculture will provide products that pose no more, and in certain instances fewer, hazards than those associated with the traditional wild-capture fisheries. Nevertheless, public health issues associated with aquaculture can be complex, and it is important to understand that, as with any technology, there can be shrouded hazards that will have either direct or future impacts and unintended consequences on human, animal, and environmental health issues. Garrett et al. (1997) categorized these issues under the general areas of technology ignorance, technology abuse, and technology neglect (Figure 3.2). Examples of these issues include (1) ignorance of the microbiological profile of aquaculture products that may affect human health; (2) abuse and willful misuse of therapeutic drugs, chemicals, and/or natural fishery habitat areas; and (3) neglect of proper training of employees in the use and application of therapeutics and/or chemi-

TABLE 3.1. Aquaculture Production of Top 20 Ranked Industrialized Countries, 1998

Rank	Country	Production (K mt)	% of Total Production	Cumulative % of Total Production
1	Japan	766.8	22.5	22.5
2	United States of America	445.1	13.0	35.5
3	Norway	408.9	12.0	47.5
4	Spain	313.5	9.2	56.7
5	France	274.9	8.0	64.7
6	Italy	246.6	7.2	72.0
7	United Kingdom	137.4	4.0	76.0
8	Netherlands	120.0	3.5	79.5
9	New Zealand	93.8	2.7	82.3
10	Canada	90.6	2.7	84.9
11	Germany	67.0	2.0	86.9
12	Russian Federation	63.2	1.9	88.7
13	Greece	59.9	1.8	90.5
14	Denmark	42.4	1.2	91.7
15	Ireland	40.4	1.2	92.9
16	Poland	29.8	0.9	93.8
17	Ukraine	28.3	0.8	94.6
18	Australia	27.8	0.8	95.4
19	Faeroe Islands	20.6	0.6	96.0
20	Israel	18.6	0.5	96.6
	Total	3411.3		

TABLE 3.2. Aquaculture Production of Top 20 Ranked Species in Industrialized Countries, 1998

Rank	Species	Production (K mt)	% of Total Production	Cumulative % of Total Production
1	Atlantic salmon	580.8	17.0	17.0
2	Blue mussel	499.9	14.7	31.7
3	Pacific cupped oyster	403.6	11.8	43.5
4	Rainbow trout	339.1	9.9	53.4
5	Channel catfish	256.0	7.5	61.0
6	Yesso scallop	226.5	6.6	67.6
7	Mediterranean mussel	158.6	4.6	72.2
8	Japanese amberjack	146.8	4.3	76.5
9	Common carp	146.8	4.3	80.9
10	Silver sea bream	82.5	2.4	83.3
11	New Zealand mussel	75.0	2.2	85.5
12	American cupped oyster	55.7	1.6	87.1
13	Carpet shells	48.0	1.4	88.5
14	Gilthead sea bream	39.3	1.2	89.7
15	Silver carp	37.3	1.1	90.8
16	European sea bass	29.0	0.9	91.6
17	Japanese eel	22.0	0.6	92.2
18	Northern quahog	19.9	0.6	92.8
19	Red swamp crawfish	17.2	0.5	93.3
20	Tilapia	16.1	0.5	93.8
	Total	3411.3		

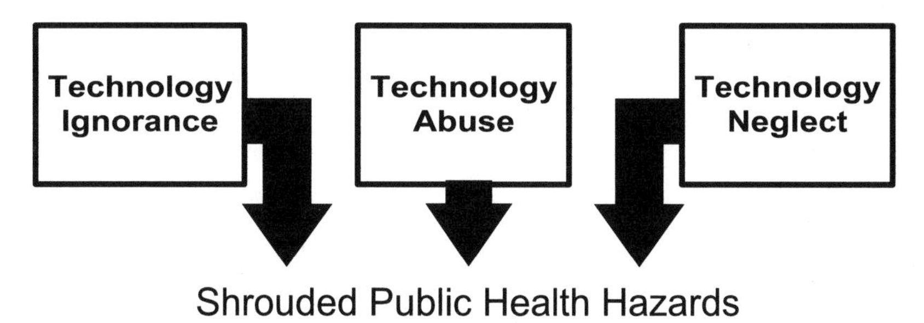

Figure 3.2. *Unintended Consequences of Aquaculture*

cals (Garrett et al. 1997). WHO (1999) published a report detailing potential food safety issues associated with aquaculture products. These issues focused on biological hazards (e.g., parasites and bacteria); and chemical hazards (e.g., agrochemicals, water treatment compounds, pesticides, chemotherapeutants, antimicrobial agents, heavy metals, organic pollutants, feed ingredients, and contaminants, etc.).

Biological Hazards

Parasites In industrialized countries parasites from aquaculture are a definite, but not major, public health problem. However, as eating habits change, and with increased trade and globalization of seafood products, parasites may become a more important problem in industrialized countries. Public health risks do exist from parasites in aquacultured products, but they are highly focused and involve products that are consumed raw or undercooked. The route of infection from parasites to humans is through the consumption of raw or undercooked seafood, which can be controlled through well-designed and -implemented HACCP programs. Consumer education programs providing information on the risks associated with the consumption of raw seafood can also be effective in reducing infections by parasites.

Organisms of public health concern include the Protozoa. Examples of these organisms include *Cryptosporidium parvum*, which may contaminate the marine environment and have been found in shellfish. Microsporidia have been found in marine mammals, which poses the issue of possible infectivity to humans. Amoebae can be a potential issue in freshwater and perhaps in estuarine environments. *Giardia* has been found in seafood harvested from Tokyo Bay contaminated with sewage (National Marine Pathogen Workshop 2000).

Helminths (worms) are also significant human health hazards in certain areas of the world in that they are parasites that cause disease in humans by being transmitted to humans via fish or crustaceans (Howgate 1997, 1998). Many of the parasites of concern belong to the classes of Nematoda, Cestoda, and Trematoda. The primary nematodes causing human diseases are *Anisakis simplex* and *Pseudoterranova dicipiens*. Normal cooking procedures and freezing (e.g., –23°C for 7 days or –35°C for 15 h) will kill these worms. However, cooking and freezing may not protect susceptible people from allergic reactions to ingested anisakids (Audicana et al. 1997). Farmed salmon in the United States, Norway, and Scotland have been shown to be free of nematodes when fed exclusively on formulated pelleted diets (Angot and Brasseur 1993; Deardorff and Kent 1989; FDA 1998). This should also be true for other marine aquacultured species if they are only fed commercially prepared pelleted feeds and not with supplemental raw fish.

Cestode infections in humans are not common and never cause fatalities in people (WHO 1999). The primary cestode of concern is *Diphyllobothrium latum*. Salmon is the most common finfish responsible for transmission of diphyllobothriasis to humans. Symptoms of infection include general body weakness, nausea, diarrhea, and pain in the abdominal area.

Trematode infections can cause fatalities in humans and are responsible for a variety of parasitic diseases throughout the world, especially in freshwater aquatic products raised in tropical and subtropical zones (WHO 1995). The trematodes *Clonorchis sinensis* and *Opisthorchis viverrini* have caused liver damage in infected individuals, and evidence indicates a possible link to liver cancer. To better understand and control infections by these species, information is needed on trematode life cycles, modes of transmission, reservoir, and vectors.

Bacteria In industrialized countries, public health issues associated with the microbial profile of aquacultured products are similar to those of wild-caught species. Howgate (1997) reported no reason to expect the risk of food poisoning from farmed marine fish or from products prepared from farmed marine fish to be greater compared with corresponding wild species. In addition, the risk from fecal organisms may be lower. There are few reports of human illnesses associated with consumption of finfish (Jensen and Greenless 1997). In addition, similar to wild species, any potential microbial safety risks can be reduced by cooking, sanitation during processing, handling and storage, and personal hygiene (Jensen and Greenless 1997). The following is a brief discussion of microbial issues and aquacultured products.

Bacteria associated with aquatic products (both wild captured and aquaculture) can be grouped into indigenous organisms naturally present in the environment, anthropogenic organisms in the environment, or those introduced into the product during posthandling and processing. Common, naturally occurring bacteria found in the aquaculture environment include members of the Vibrionaceae (e.g., *Vibrio vulnificus*, *V. parahaemolyticus*, *V. mimicus*, *V. hollisae*, *V. cholerae*), *Clostridium botulinum* Type E, *Listeria monocytogenes*, *Plesiomonas shigelloides* (formerly *Aeromonas shigelloides*), and *Mycobacterium* spp. (Hackney and Dicharry 1988; Wolfe 1998). *Edwardsiella tarda*, *Aeromonas hydrophila*, and *P. shigelloides* have been isolated in catfish ponds but are generally more commonly found in freshwater compared with marine or brackish water aquaculture systems (Leung et al. 1992; Wyatt et al. 1979b). *C. botulinum* has been isolated in fish and sediments in freshwater trout farms in Britain and Denmark (Cann et al. 1975; Huss et al. 1974). King and Flick (2000) isolated *A. hydrophilia*, *P. shigelloides*, *Vibrio* spp., *Yersinia enterocolitica*, and *Bacillus cereus* from six freshwater and two mariculture recirculating aquaculture facilities.

Anthropogenic organisms include *Salmonella*, *Shigella* spp., etc., which have been found in aquaculture production sites (Wyatt et al. 1979a). Rodents, insects, birds, employees, contaminated water supplies, etc., can spread these organisms throughout the facility. There are also fish pathogens capable of causing disease in humans. A partial survey of these zoonotic bacteria includes *Aeromonas* spp., *E. tarda*, *Erysepilothrix rhusiopathae*, *Mycobacterium* spp., *P. shigelloides*, *Streptococcus iniae*, and *Vibrio* spp. (Wolfe 1998). *Leptospira* spp. can be transmitted to aquaculture systems through the urine of rodents.

The high stocking densities associated with aquaculture provide an optimum environment for pathogen growth and survival (Jahncke and Schwarz 1998). Bacteria can also be transmitted to people who come into contact with the production systems (e.g., *Vibrio* infections through open wounds or cuts) (Blake et al. 1997). Control of these organisms, within aquaculture facilities, requires the development of standard operating procedures (SOPs), implementation of good manufacturing practices (GMPs) and best management practices (BMPs) in conjunction with employee training programs, and application of HACCP principles as a risk management approach (Jahncke and Schwarz 1998; 2000).

Potential public health issues associated with the processing and distribution of aquaculture products are similar to those found in wild-caught species. These issues can be categorized as process induced, distribution induced, or consumer induced (Garrett et al. 1997). Processing hazards of aquacultured products often include organisms such as *C. botulinum* and *C. perfringens*, *Staphylococcus aureus*, *L. monocytogenes*, *Salmonella* spp., *Shigella* spp., *Yersinia* spp., and *Campylobacter* spp. (Garrett et al. 1997). Distribution hazards of these products generally relate to improper temperature control with the potential for proliferation of pathogenic microorganisms. Consumer hazards relate to food acquisition, food handling, food preparation, and serving and storage of leftovers (USDA 1989). For each of the years from 1993 through 1997, the most commonly reported food preparation practice for all foods that contributed to foodborne disease was improper holding temperatures. The second most commonly reported practice was inadequate cooking (CDC 2000).

As with wild-caught species, the microbial profile of harvested aquacultured products can affect human health, as evidenced by transmission of streptococcal infections from tilapia to humans in Canadian fish processors (CDC 1996; Weinstein et al. 1997). Similarly, the transmission of *Vibrio* spp. infections from aquatic animals to workers in Israel was a result of changing the distribution pattern of marketing processed fish to marketing live fish (Bisharat and Raz 1996). Food handling and food processing hazards can be addressed by using appropriate food safety measures such as HACCP principles (WHO 1999).

Viruses Viruses causing diseases in aquatic organisms are not pathogenic to humans (WHO 1999), and transmission of aquaculture viruses through wastewater reuse systems is of little concern to public health. Nevertheless, bivalve mollusk filter feeders can concentrate human pathogens such as hepatitis A and E and Norwalk-like viruses. Consumption of raw molluscan shellfish is the leading cause of viral illnesses in people who consume seafood products (Jensen and Greenless 1997; Garrett et al. 1998; Vaughn et al. 1984).

Jensen and Greenless (1997) indicated that the most challenging public health risks associated with aquaculture were with shellfish grown in open surface waters that may contain both naturally occurring and anthropogenic contaminants that can lead to disease outbreaks. Jensen and Greenless (1997) stressed that aquaculture operations can minimize public health risks by ensuring proper site evaluations and following good agricultural practices (GAPs). Shellfish management programs to control public health risks associated with shellfish aquaculture include strict acceptance of public health classifications and monitoring of growing waters, proper siting of aquaculture areas, enforcement of BMPs, and training of employees and harvesters in sanitation procedures. Additionally, public education programs are needed to inform consumers about the potential for contracting viral diseases from consuming raw molluscan shellfish products.

Nevertheless, although such management programs will help to reduce they will not eliminate the public health risks associated with consumption of molluscan shellfish. For example, in the United States, despite a well-established shellfish monitoring program, consumption of molluscan shellfish is one of the leading causes of seafood-borne illness (CDC 2000). Classification of harvesting waters and numbers of indicator bacteria found in growing waters are not an absolute gauge for determining the presence or absence of pathogenic bacteria and viruses in the environment. Fecal coliform indicator bacteria do not correlate well with the presence of naturally occurring pathogenic *Vibrio* spp. and with human enteric viruses that may be present in growing waters (Cole et al. 1986; Elliot and Colwell 1985; Kilgen et al. 1988; Paille et al. 1986; Richards 1988). In addition, although depuration has been used to help remove enteric pathogenic bacteria from molluscan shellfish, there are differences in the rate of depuration for various enteric viruses and marine *Vibrio* spp. (Ahmed 1991; Rose and Sobsey 1993). For example, Abad et al. (1997) reported that shellfish still contained rotavirus and hepatitis A viruses after 96h in a flow-through ozonated marine depuration system. A more effective treatment approach may be the use of postprocess methods such as high hydrostatic pressure (HHP) and irradiation which have been shown to be effective against pathogenic bacteria. However, additional research is needed on these procedures to determine specific treatment parameters needed to eliminate pathogenic viruses from molluscan shellfish products.

Chemical Hazards

Chemical concerns of aquacultured products include natural biotoxins such as those that can cause paralytic shellfish poisoning (PSP), amnesic shellfish poisoning (ASP), neurotoxic shellfish poisoning (NSP), and diarrhetic shellfish poisoning (DSP), etc. Public health protection is based on careful site selection and monitoring and surveillance programs for known toxin-producing algae. Other programs such as public education and water classification programs can also protect public health (Boesch et al. 1997; Jensen and Greenless 1997).

Additional chemicals of concern include organochlorine pesticides such as aldrin, endrin, dieldrin, chlordane, toxaphene, chlordecone, dichlorodiphenyltrichloroethane (DDT), diphenylethanedichlorophenyl (ethane) (TDE), dichlorodiphenyldichloroethylene (DDE), heptachlor epoxide, mirex, PCBs, and dioxins. However, proper site selection for aquaculture operations can minimize the risk to aquacultured species by contamination from organochlorines. Global databases identifying contaminated areas are currently being developed (Garret et al. 1998). These databases, coupled with new seafood consumption studies, will provide additional food safety protection for consumers that will allow for a more innovative approach to risk assessments compared with traditional methods that rely on historical information.

However, contamination problems can still occur. In 1997 some animal feeds, including pelleted diets used in the US's catfish industry, were contaminated with dioxin. The US Food and Drug Administration (USFDA) and state agencies quickly identified the problem and traced the dioxin contamination to ball clay used in the feeds. Catfish fed with the contaminated feeds were analyzed. The analyses showed low concentrations of dioxin in the edible portions of the fish that were not significantly different from tissue background concentrations in other farmed catfish (Fielder 1998; Rappe et al. 1998).

Pesticides and chemicals used in aquaculture must be approved by regulatory agencies for use on food fish at aquatic production sites. Other aquaculture chemicals (e.g., lime, potassium permanganate, hydrogen peroxide, calcium hypochlorite, aluminum sulfate, ferric chloride, calcium sulfate, zeolite, etc.), when used according to GMPs, are not considered hazardous to consumers (FAO 1997).

Heavy metals are also a public health concern but are rarely a problem for aquacultured organisms because high-risk areas can be avoided for growing purposes. A recent study showed concentrations of metals and pesticides in farm-raised catfish, rainbow trout, and crawfish well below USFDA or US Environmental Protection Agency (USEPA) regulations (Food Chem News 2001). On the other hand, wild species may have a greater risk of trace metal accumulation compared with aquaculture species, because there are no controls to prevent wild species from inhabiting locations contaminated with chemicals and heavy metals (Jensen and Greenless 1997).

Aquaculture Drugs Disease outbreaks within an aquaculture operation can be a common occurrence that, if unchecked, as with any animal husbandry endeavor, can rapidly destroy an entire crop. In the United States, the use of antibiotics and drugs is regulated by the USFDA (JSA 1994). These compounds can only be used for certain species and life stages. There also are designated withdrawal periods during which the crop cannot be harvested and sold. Similarly, in other countries, antibiotics and other compounds are available for use only with a veterinary prescription (Bangan et al. 1994; 1996).

The European Union (EU) is implementing a program to test routinely for a range of veterinary drug compounds in aquacultured products. This program, along with the increased harmonization of international food safety standards for aquacultured products, will help ensure safety as regional monitoring and testing programs become more common (WHO 1999). Testing and monitoring protocols and established withdrawal periods will help ensure that no harmful residues remain in the edible tissues of these products (WHO 1999).

Nevertheless, the concern in antibiotic use is the possible development of drug-resistant human pathogens (D'Aoust 1994; Garrett et al. 1997). It is generally recognized that as more antimicrobials are used there will be a higher frequency of resistant microorganisms (WHO 1999). However, it is difficult to evaluate this assumption, because there is little agreement on how to define resistance in pathogenic bacteria and no consensus on how to measure resistance (WHO 1999).

Smith et al. (1994) reported that there is a tendency to emphasize potential risks of resistant microorganisms instead of objectively assessing their importance to public health. There is evidence that antibiotic resistance in bacteria can be selected during normal therapeutic use. The risks of transfer of resistance to human consumers is low (Alderman and Hastings 1998), and the reputed risk to public health is probably limited to indirect exposure to antimicrobials (WHO 1999). Similarly, Howgate (1997) stated that evidence for any possible harm to humans from the use of veterinary drugs in aquaculture is difficult to substantiate.

However, bacteria resistant to antimicrobial agents have been isolated in sediments beneath net pens (Kerry et al. 1994). Resistant bacteria have also been isolated in the intestines of commercial fish in net pens (Ervik et al. 1994). Antibiotic resistance in bacteria is carried on R plasmids (Watanabe et al. 1977; Aoki 1997). Horizontal transfer of resistant genes has occurred between bacteria in aquaculture pond water and in aquaculture pond sediments (Aoki 1997; Steward and Sinigalliano 1990). In addition, researchers have successfully demonstrated in vitro transfer of antibiotic resistant genes between fish and human pathogens (Hayashi et al. 1982; Nakajima et al. 1983; Sandaa et al. 1992; Son et al. 1997).

According to Jensen and Greenless (1997), the key to reducing drug use in the aquaculture industry, and thus to reduce potential public health risks, is to integrate the proper use of drug application in conjunction with GAPs. For example, although production levels have increased in Norway, antibiotic use has declined dramatically (Hatt 1998a; WHO 1999). In 1987, 50,000 kg of antimicrobial agents were used in Norwegian aquaculture operations. Since that time, aquaculture management plans have been implemented that incorporated routine vaccinations for fish, aquaculture site rotations, and separation of different fish generations. These efforts have minimized disease outbreaks and reduced the antimicrobial agents to approximately 746.5 kg/year (Hatt 1998a). In addition to these efforts, it is recommended that aquaculture employees receive training on how to properly use, store, and dispose of medicated feeds, chemotherapeutants, and chemicals (Jahncke and Schwarz 1998). Inappropriate use and/or misuse of such compounds are not only illegal but can pose potential health risks to employees and to the public (Garrett et al. 1997).

ANIMAL HEALTH ISSUES

Diseases from wild stocks may infect new aquaculture species, or, conversely, diseases from aquacultured species may infect wild stocks. Risk management plans are needed to control these possible epizootic animal health issues. Countries around the world that have neglected such disease control procedures have suffered negative consequences to their fisheries. Farmed fish and shellfish have been implicated as reservoirs of disease organisms (Monro and Waddell 1984). Weston (1991) reviewed the available scientific literature and reported that 48 parasites of freshwater fish were transferred around the world via importation of live fish.

Biological Hazards

Parasites Various groups of parasites (e.g., protozoa, helminths, crustaceans) can be associated with aquatic organisms. A better understanding of their life cycles, life histories, and host-parasite associations is imperative for their control. Aquaculture conditions (e.g., high stocking densities, environmental stress, etc.) can increase the likelihood of parasites multiplying to the point where they impact the health of the cultured and wild stocks. Stock enhancement programs may provide a route for transmission of parasites into new environmental settings. Similarly, wild stocks can be a source of parasites for aquacultured species (National Marine Pathogen Workshop 2000).

Life cycles of marine parasites are generally more complex and thus more difficult to control than those of freshwater fish. Eventual control of marine parasites in aquaculture production will rely on better understanding of their life cycles chemical controls, as well as biological controls, quarantine, and early diagnosis.

In Europe and North America, parasitic sea lice (*Lepeophtheirus salmonis* and *Caligus elongatus*) are a significant economic problem to the net pen salmon farming industry (Pike and Wadsworth 1999). In eastern Canada and the state of Maine, USA, sea lice are a major cause of death and loss of revenues in the Atlantic salmon industry. In 1995, salmon losses due to sea lice were estimated at more than 20% of market value (MacKinnon 1997). Sea lice are estimated to cost the Scottish industry approximately US$28.8 million to US$43.2 million annually (Sinnot 2000).

Control of sea lice infestations are currently accomplished by using chemical, biological, and fallowing techniques. Development of an effective vaccine is also underway (MacKinnon 1997, Roth 2000). Chemicals used to control sea lice include: oxidizers—hydrogen peroxide; organophosphates—dichlorvos and azamethiphos; benzoylphenyl ureas—teflubenzuron and diflubenzuron; avermectines—ivermectin, emamectin, and doramectin; and pyrethrin/pryrethroid compounds—pyrethrum, cypermethrin, and deltamethrin (Mackinnon 1997). Nine of these compounds are available for use in Norway, six are available in the United Kingdom, four in Canada, and only two are available in the United States. The use of dichlorvos is being phased out in Norway because of adverse affects to the environment and aquatic species.

Sea lice can also pose a health threat to surrounding wild fishery stocks. The Baltic and Nordic News Service (1999) reported that scientists found 90% mortality in wild smelt which were infected with sea lice transferred from a net pen salmon-rearing operation. Sea lice are also a significant threat to wild salmon stocks. However, through the use of vaccines, application of BMPs, and the use of therapeutants impacts to wild stocks are being reduced.

Chemicals also are being utilized to control sea lice infestations, although most countries require some form of an environmental risk assessment to be conducted before use of these compounds. In addition to chemicals and vaccines, several species of wrasse (*Crenilabrus melops*, *Ctenolabrus rupestris*, and *Centrolabrus exoletus*) are being used in Europe to control sea lice. However, there is evidence that these wrasse may be susceptible to infections by *Aeromonas salmonicida*, thus raising concerns that these fish, in turn, could act as a reservoir for infections (WGMAFC 2000). Fallowing techniques that incorporate sequentially rotating net pen sites that have been free of pens for predetermined periods of time are used to help reduce disease levels during periods of nonuse.

Myxobolus cerebralis (whirling disease) is an example of a myxosporan parasite that has detrimental impacts on aquacultured as well as natural fish populations. Whirling disease affects trout stocks throughout Europe. It was in the state of Pennsylvania, USA, in 1956, and has since spread to most regions of the United States through trout populations (Tamplin 1997). In wild, otherwise healthy populations, the disease is often sublethal; however, hatchery infections often result in extensive mortalities, with surviving fish acting as reservoirs of the immature stages of this parasite. Unlike sea lice, there are no therapeutic remedies available for this parasite. Management protocols to decrease intensity of infection in hatchery fish are mandatory for its control. Because there is no known effective treatment for larger fish in production, control is effected on a case-by-case basis to determine whether destruction is the best option.

Examples of parasites that affect molluscs include *Bonamia* spp. (an intrahemocytic parasite of the blood), *Haplosporidium* spp. (MSX), *Marteilia* spp., and *Perkinsus* spp. (a histozoic parasite) (Stickney 2000). *Perkinsus* spp. complexes (e.g., *P. atlanticus* and *P. marinus*) are parasites responsible for high mortalities in clams and oysters. *P. marinus* (Dermo) is at or near 100% prevalence in *C. virginica* from the USA Gulf of Mexico region to the state of Massachusetts, USA. *Haplosporidium* spp. (e.g., *H. nelsoni* [msx] *H. costale*) are also responsible for shellfish mortalities. Control and management methods for these parasites include development of parasite-resistant molluscan species and the transfer of nonnative sterile, disease-resistant oyster species into areas where parasites such as Dermo and MSX are endemic. This latter approach is under consideration in the Chesapeake Bay area of the United States, where MSX and Dermo have decimated native oyster stocks.

Bacteria Diseases are spread either vertically (i.e., from parents to offspring through sex cells) or horizontally (i.e., from one organism to another via direct contact, air, or water) (Nicholson 2000). A major difference between disease transmission in aquaculture and terrestrial animals, however, is that the water itself may promote the spread of the disease-causing agents. Bacteria can be divided into Gram-negative and Gram-positive groups. Major disease causing Gram-negative bacteria families include Enterobacteriaceae, Aeromonadaceae, Vibrionaceae, and Pseudomonadaceae (Hawke 2000).

Edwardsiella septicemia (Enterobacteriaceae) has been reported in North America, Europe, Japan, and Australia, occurring most commonly in channel catfish, common carp (*Cyprinus carpio*), striped bass (*Morone saxatilis*), red sea bream (*Pagrus major*), Japanese flounder (*Paralichthys olivaceus*), ayu (*Plecoglossus altivelis*), tilapia (*Oreochromis* spp.), and yellowtail (*Seriola quinqueradiata*) (Hawke 2000). Enteric septicemia of catfish (ESC) has been identified as the most significant disease of cultured channel catfish in the United States (Plumb 1999). It also has considerable international significance (Anon. 1995).

Furunculosis (Aeromonadaceae) has been responsible for diseases in Atlantic salmon and rainbow trout in Canada, the United States, France, Germany, Italy, and Norway. Coldwater vibriosis (CV) affects primarily the cultured salmon operations in Norway. It has also been found in the Shetland Islands of Scotland, the Faeroe Islands, eastern Canada, and the eastern United States (Plumb 1999). Saltwater columnaris (aerobic Gram-negative rods) has been found in sea bream (*Sparus* spp.) and flounder (*Paralichthys olivaceus*) in Japan and has been isolated from Dover sole (*Solea solea*) in Scotland (Hawke 2000). It is responsible for mortalities in sea bass (*Dicentrarchus labrax*) in France (Bernardet et al. 1994) and turbot (*Pleuronichthys*), Atlantic coho, and Chinook salmon (*Salmo* spp.) in Spain (Pazos et al. 1996).

Major disease-causing Gram-positive bacteria families include Streptococcaceae, *Streptococcus*-like bacteria, aerobic and anaerobic Gram-positive rods, and Gram-positive acid fast rods (Hawke 2000). Streptococcosis (Streptococcaceae) is a common disease associated with aquacultured fish, causing lesions and mortalities. Often, this pathogen can remain as a sublethal nuisance until environmental parameters are exceeded and resultant fish stress allows for opportunistic infections. *Streptococcus* spp. is a pathogen of increasing concern in warm-water RAS culture. Mycobacteriosis (Gram-positive rods) has been found in wild striped bass (*Morone saxatilis*) in the Chesapeake Bay and Maine, USA and is a health concern for fish handlers.

Bacterial kidney disease (BKD) (aerobic Gram-positive rods) is an infectious disease of salmon with escaped aquacultured fish serving as potential reservoirs of infection for wild stocks. BKD is quickly transmitted both horizontally and vertically to fish raised under high densities. It is reported that 15–20% of wild stocks of salmon are infected with BKD (Tamplin 1997).

The current treatment for bacterial pathogens generally incorporates the use of antibiotics. Administration of these drugs can be as an immersion for small fish, in feed for larger fish, and injections for brood stock. Vaccinations can also be effective to control pathogens, but this approach must be monitored carefully. For example, the impact of *Aeromonas* spp. on aquaculture has diminished through utilization of vaccinations; however, evidence is surfacing that antibiotic-resistant strains can be transferred from aquacultured stocks to wild stocks.

A combination of BMPs, development of new vaccines, wise use of approved drugs, and good facility management protocols may be the best approach to control diseases. In that regard, many countries are in the process of implementing and establishing health management programs that require periodic testing and certification of health status of aquaculture organisms (Nicholson 2000).

Viruses Viral diseases can be devastating to aquaculture operations, and they pose a potential threat to wild stocks as well. New viruses are being discovered as new species are being introduced around the world. Viruses can survive for extended periods of time in water and sediments and can be easily transmitted within species groups.

A few examples of major virus families that infect finfish include Birnaviridae, Herpesviridae, Orthomyxoviridae, and Rhabdoviridae (Brady 2000). Infectious pancreatic necrosis (IPN) (Birnaviridae) is manageable in salmon culture but can cause high mortalities in halibut. In 1998 viral hemorrhagic septicemia (VHS) was reported in a rainbow trout hatchery in western Norway. This was the first report in rainbow trout in Norway since 1974, and the strain was similar to one isolated from rainbow trout in Denmark (ICES 1999).

Channel catfish (*Ictalurus punctatus*) virus disease (CCVD) (Herpesviridae) is responsible for significant economic losses in areas where channel catfish is cultured. Fingerlings less than eight inches in length are primarily affected whenever high temperatures and secondary stressors (i.e., extensive handling, parasites, low dissolved oxygen, etc.) are present (Tucker 2000). Considerable research efforts have been directed to developing vaccines for CCVD. Nevertheless, an economically viable vaccine has yet to be developed. Currently, CCVD is primarily a concern of the United States, as European wild catfish are not prone to this disease.

Infectious salmon anemia (ISA) (Orthomyxoviridae) is a major concern for salmon operations and has been a major problem in Norway during the past 15 years. ISA was in eastern Canada in 1997 and in Scotland during 1998. It is responsible for major financial losses and is a serious threat to the continued success of salmon farming. The ISA problem in New Brunswick, Canada is reported to have affected the overall production of Atlantic salmon, and the disease is the focus of ongoing debates of impacts of ISA from cultured to wild stocks

(ICES 1999). Confirmation of ISA in escaped Atlantic salmon has been documented. However, the cause of ISA found in wild salmon from the Magaguadavic River in eastern Canada is under investigation. It was recently isolated from net pen salmon in the state of Maine, USA (Anon. 2001). In Scotland all outbreaks of ISA were linked to a single facility. The spread of ISA is closely linked to transferring live fish between farms, transfer of the virus via contaminated equipment, and untreated salmon processing plant wastes (ICES 1999).

DNA viruses associated with marine fish culture include lymphnocystis disease, red sea bream iridoviral disease (RSIVD), viral epidermal hyperplasia (VEH), and salmonid herpes virus type 2 infection (Nakajima 1997). RNA viruses associated with marine fish culture include yellowtail ascites virus (YAV), viral deformity virus (VDV), hirame birnavirus infection, Rhabdovirus, and virus nervous necrosis (VNN) (Nakajima 1997). Other viral diseases of finfish include erythrocytic inclusion body syndrome (EIBS) and Kuchijiro-shou (snout ulcer disease).

Viral diseases can also affect the health of crustaceans. Major families of viruses that infect crustaceans include Baculoviridae, Picornaviridae, and Parvoviridae. Recent episodes in shrimp culture include the introduction of Taura syndrome virus (TSV) (Picornaviridae), Monodon baculovirus, white spot syndrome baculovirus complex (WSSV), and yellowhead virus disease (YHV) (Baculoviridae) into shrimp aquaculture facilities. Infectious hypodermal and hematopoietic necrosis virus (IHHNV) has been found in culture facilities and in wild stocks in North America, South America, and Asia (Brady 2000). TSV was identified in outbreaks in the states of Hawaii, Texas, South Carolina, USA, and Central and South America. Studies in the United States suggest that native white shrimp may also be susceptible to these viruses (Lightner 1999). The potential for disease transfer, specifically shrimp viruses, into native wild stocks has been recognized by the Texas Parks and Wildlife Department, USA as the single most serious resource management issue (McKinney 1997).

Vaccinations and health screening of animals can help control viral diseases in aquaculture. Disease-resistant species must be developed in conjunction with implementation of risk management programs to control viruses at production and processing sites. Databases on disease outbreaks should be collated and reviewed to determine possible disease links throughout the world. In addition, establishment of aquatic fish health programs that include periodic testing of aquaculture species for diseases, development of pathogen-free eggs, and larvae and brood stock programs is needed. Aquatic health programs will be more effective in controlling diseases as more sensitive, rapid, and quantitative disease diagnostic tests are developed.

Entry of exotic pathogens into aquaculture operations can occur via contaminated feeds, infected brood stock, eggs and larvae, birds, water insects, employees, equipment, etc. Threats to wild stocks can occur via effluent waters, aquaculture waste products, the escape of infected organisms, processing plant wastes, and improper waste disposal methods. Efforts to manage exotic pathogens should focus on the development and implementation of biosecure facilities, vaccines, specific pathogen-free (SPF) stock, and the use of HACCP principles as a risk management approach at the production and processing facilities.

ENVIRONMENTAL HEALTH ISSUES

Environmental groups are concerned about the potential negative impact of aquaculture operations. These concerns center around issues such as aquaculture effluent and its effect on inland and coastal waters, escapement of nonnative species, and habitat destruction (Emerson 1999). Other concerns include spread of diseases, the use of chemotherapeutants, cage and pen site densities, and use of nonnative species.

Folke et al. (1998) suggest that an estimate of the ecological footprint resulting from proposed aquaculture operations should be an important component of the siting and development of culture operations. Hankins (2000) suggested that the aquaculture community take a more active role in developing its own environment by taking a lead role on the issues of nutrient loading, water quality, and environmental stewardship. New approaches and technology are needed to address eutrophication impacts on the environment while maintaining or increasing operational profitability, and finally, to develop more biosecure production facilities (Hankins 2000). Many countries are implementing environmental guidelines for aquaculture. For example, the USEPA has initiated a rule-making process to require BMPs for aquaculture effluents (Boyd 2000).

Biological

Aquaculture management safeguards are needed to help protect the environment. Such management approaches should include the development of detailed operating procedures to address disease issues including equipment cleaning protocols and proper handling of waste stream discharges (Raynor 2000). A few examples follow.

A Scottish aquaculture management plan addresses the protection of water quality and organic enrichment and nutrient releases from the pens. It also addresses maximum biomass and discharges of nutrients, licensed medicines, or unauthorized chemical formulations. This plan is being developed in response to public pressure regarding environmental impacts and outbreaks of ISA (Inglis 2000). Scotland is also initiating studies to identify the benthic impact of shellfish farming. In France, similar studies are being conducted to determine impacts of biodeposits from oyster culture areas.

The aquaculture industry in Victoria, British Columbia, Canada, will provide data to environmental regulators on environmental conditions around net cages. This information will allow their regulatory agencies to set future environmental standards to help determine which sites will require restoration, relocation, or both. The goal of these standards is to provide information to help the industry properly site its operations as well as to provide assurances to the public concerning environmental and animal health issues (Anon. 2000).

The Japanese are currently implementing regulations to improve and preserve aquaculture grounds and to prevent the spread of fish and shellfish diseases by developing guidelines to monitor, detect, and then properly dispose infected stocks (Morikawa 2000).

Possible interbreeding between escaped cultured and wild species and displacement of native species by escaped nonnative cultured species is also a significant environmental concern. Protection of indigenous stocks is a high priority in many countries. The US National Marine Fisheries Service (USNMFS), part of the US Department of Commerce (USDOC) recently listed wild populations of Atlantic salmon in rivers and streams in the state of Maine from the lower Kennebec River to the United States–Canadian border as endangered (NMFS 2000b). In that regard, net pen salmon culture may be eliminated in these areas unless salmon farmers can provide assurances that escaped cultured salmon will not interbreed with wild stocks.

Nutrients

The recycling of nutrients using integrated culture methods is environmentally sound. Pond aquaculture operations in industrialized countries are investigating the use of closed-system ponds with little water discharge in combination with settling ponds and bivalve and artificial wetland programs to reduce deleterious materials in aquaculture discharge waters (Samocha et al. 1997). Denitrification filters can accomplish nitrate removal from discharge waters. Phosphorus removal, on the other hand, can be expensive; thus the development of feeds with higher phosphorus digestibility should be a high priority (Golz et al. 1997). A combination of hydroponic vegetable gardens and wetlands have shown promise to significantly reduce phosphorus and nitrogen discharges from aquaculture operations. In Israel, gray mullet (*Mugil cephalus*) have been raised in cages placed below the cage culture of sea bream (*Sparus aureate*) and red drum (*Scienops ocelatus*) to consume particles from the cages containing sea bream and red drum. Such polyculture was shown to help reduce particulate matter in the pond discharge water (Hatt 1998a).

Modern feeds should be designed to maximize their nutrient absorption by the aquacultured animals and thereby minimize losses of feed materials and nutrients into the environment. Feed and feeding unit operations are expensive; thus aquaculture firms have considerable incentives to maximize the use of environmentally sound foods and feeds. In addition, the recent development of cages and shellfish culture systems designed for open ocean areas can reduce the impact on protected inshore sites.

The USEPA is currently developing pollution controls for nationally applied discharge standards (i.e., effluent limitations, guidelines, and standards) for commercial and public aquatic animal production facilities (Federal Register Notice 2000). To date, in the United States there are no nationally applicable effluent limitations or guidelines and standards to regulate discharges from aquaculture facilities and there is no unified federal or state regulatory framework for marine aquaculture (DeVoe 1997).

Canada is initiating research on possible impacts on the benthic environment from shellfish farming operations. Data being collected include settling rates of feces and pseudofeces with a variation of seston levels and incorporating benthic primary production components. The Nova Scotia School of Fisheries and Aquaculture is testing a unique application of International Standard (ISO) 14000 to a working fish farm in southwest Nova Scotia (Hatt 1998b). The standards will focus on environmental policy to help implement and operate an environmental monitoring system (EMS), including periodic reviews of the overall EMS. Adherence to the standards has helped to reduce effluent wastes from the facility.

Norway and Scotland developed the first environmental monitoring programs for finfish in the early 1990s. A Norwegian management plan for aquaculture addresses sedimentation rates, chemical condition of sediments, and benthic infauna communities. Management approaches focus on determining the local environmental impact on holding capacities of the sites. Regulations have been developed limiting the levels of nitrogen, phosphate, and particulates that can be released into the environment (ICES 1999).

France has implemented annual checks and sampling of aquaculture operations by the operator and by regional authorities. A process was implemented for the French Ministry for Agriculture and Fisheries to determine fitness zones to identify suitable locations for shellfish and fish aquaculture firms.

Ireland has implemented a monitoring plan for salmon farms to determine whether additional production capacity should be allowed at a site. This plan focuses on the following three levels: (1) direct observations, (2) redox measurements of the sediments, and (3) monitoring of macroinvertebrate fauna numbers.

SOCIAL ISSUES

A combination of environmentally sound integrated farming methods, proper siting of facilities, and implementation of codes of practice and BMPs can help to minimize negative impacts of aquaculture. The emphasis on environmentally and socially friendly aquaculture is leading aquaculturists, planners, decision makers, and investors to place aquaculture in a larger context of total resource use and accountability (FAO 1997).

Codes of practice and BMPs are being developed throughout the world to address both social and environmental issues. For example, the FAO Codex Code of Conduct for Responsible Fisheries is global in scope and addresses capture, processing, and trade of fish and fishery products, fishing operations, aquaculture, research, and integration of fisheries in coastal management (FAO 1995). The Federation of European Aquaculture Producers (FEAP) is promoting a Code of Practice for the responsible development and management of European aquaculture. This organization promotes a high standard of food quality, while respecting environmental issues (Aquaculture.com 2000). This Code of Practice addresses all aspects of aquaculture, including social and economic relationships and consumers.

BMPs can be used to reduce adverse environmental impacts from aquaculture operations. BMPs are also useful in helping to improve operational aspects of a facility and to assist in identifying and addressing issues of concern before they become an environmental or community problem. BMPs are commonly used to address aquaculture waste effluent issues (e.g., effluent volumes, nutrients, total solids, oxygen demands) (Boyd 2001).

Environmental programs to monitor aquaculture operations are important because they provide surveillance data to regulatory agencies and information to the public on possible environmental impacts of aquaculture operations. Monitoring programs are being implemented in several countries for use in net pens and shellfish growing operations. Socially and environmentally friendly aquaculture is a viable approach to help promote and support community development. Aquaculture can aid in diversifying the economy of rural and coastal communities alike. Additional jobs and new opportunities offered by aquaculture will help to revitalize communities. However, for this to occur, careful assessments of regulatory and institutional constraints must be undertaken to effectively use all available resources (Skladany and Bailey 1994).

CASE STUDIES

The following two case studies are examples of aquaculture companies that are addressing the public, animal, and environmental health issues associated with their operations. These studies represent two successful and profitable companies that have established environmentally and socially friendly businesses.

Hybrid Striped Bass—California, USA

This North American case study is of Kent SeaTech Corporation, a hybrid striped bass (HSB) production facility. Key personnel began research in 1968, focusing on development of spawning techniques for finfish, crustaceans, and bivalves. From the beginning, Kent SeaTech worked closely with regulatory agencies on issues of importance to the aquaculture industry (e.g., new regulations and law, brood stock acquisition, larval rearing, water treatment, nutrition, disease).

In 1972, Kent SeaTech was incorporated and research began on tank culture of HSB, a cross between striped bass (*Morone saxatilis*) and the freshwater white bass (*Morone chrysops*). In 1972, the sale of HSB was illegal in the state of California, USA because of poaching pressure on the state's recreational fishery. By 1983 this perceived threat and other regulatory issues concerning the aquaculture and sale of HSB in California were resolved through cooperative adoption of new regulations by the California Department of Fish and Game.

In 1983, a decision was made to build a production facility in the productive agricultural Coachella Valley near Palm Springs, California, USA. In the Coachella Valley land was zoned for agriculture, geothermal groundwater was available with an average temperature of 28°C, and there were fewer multiple-user conflicts. As the production facility expanded, a cooperative water leveraging program utilizing multiple reuse of water within the facility and with surrounding agricultural farms was implemented. This program is still in use today. Implementation of this cooperative water conservation program significantly reduced reliance on well water and decreased environmental concerns regarding nutrient discharge by using these discharges on agricultural row crops. In 1984, using circular production tanks, Kent SeaTech produced 181 mt/year of HSB.

In 1985, HSB fingerling production was outsourced to another facility to streamline operations and focus on growout production. During 1988, production was doubled to approximately 454 mt/year. In 1991, a major marketing initiative was implemented and production tripled to approximately 1,361 mt/year. The increased water leveraging program allowed expansion without increasing groundwater requirements. Kent SeaTech's impressive success is due to central planning efforts and strong cooperate programs between Kent SeaTech, universities, and governmental agencies.

Public Health Issues Kent SeaTech's food safety issues are addressed by a HACCP program that focuses on possible contaminants in the fish, in the feed, and in the production system. The company has implemented HACCP principles throughout all aspects of their operation. The water supplies, fish, and feeds are analyzed regularly for possible contaminants. A fish pathologist is stationed on-site to monitor fish health daily. This proactive fish health program, combined with the development of an innovative vaccination procedure, has reduced the incidence and severity of disease and has decreased the use of USFDA-approved chemotherapeutic agents.

In addition to food safety, fish quality is also a high priority for the company. Fish are harvested and placed directly into chill-kill ice tubs to achieve a rapid reduction in flesh temperature. The tubs are transported to a state-of-the-art refrigerated packing plant, packaged with gel packs, held briefly in a refrigerated room, and then transported in refrigerated trucks to market destinations.

Animal Health Issues Animal health issues are addressed through nutrition, genetics, and a proactive fish diagnostics program. Ongoing studies on fish nutrition, in conjunction with scientists from Texas A&M University, are designed to maximize fish health and enhance fish growth and feeding efficiencies. The fish production systems are continuously monitored for pH and oxygen levels, etc., to maintain optimal water quality parameters to maximize fish growth while reducing stress in the fish. A proactive fish diagnostic program also identifies potential problems before widespread disease outbreaks occur. This program has reduced the use of chemotherapeutants. In addition, a genetics program, initiated with scientists from Texas A&M, is being used to select for more disease-resistant HSB, thus further reducing the reliance on chemotherapeutants for disease control.

Environmental Health Issues The water leveraging program that Kent SeaTech employs is a holistic approach that uses tilapia to reduce effluent solid levels and oxidation biofiltration lagoons in combination with artificial wetlands to reoxygenate the water and reduce nitrogen and phosphorus concentrations in the water for optimal reuse purposes. Aquaculture water recirculation rates are in excess of 85%. Aquaculture discharge effluents are also used to grow agricultural row crops. Thus water is reused and conserved, and the high-nutrient aquaculture effluent water reduces the amount of chemical fertilizer applied to the row crops. Not only is Kent SeaTech the largest HSB producer in the world, it is also one of the world's most environmentally friendly.

Salmon Net Pens—Ireland

The following case study was summarized from the environmental assessment written by Ecoserve (ECS), Ltd. (2000).

This western Europe case study is based on Hibernor Atlantic Salmon Ltd. located on Lough Allen, County Leitrim, Ireland. Lough Allen is 13 km in length and 3500 ha in area and is located on the upper section of the River Shannon catchment (ECS 2000). The Lough Allen has a surface area of 35 km^2, with a 31-m maximum depth (4–5 m average depth). The surrounding topography is comprised primarily of carboniferous limestone with some shale and sandstone. Soils are primarily gleys and peat, which are poorly drained. This county is sparsely populated and currently has little tourist trade, although its scenic lakes, fishing, and mountains have good future potential for increased tourism.

The site of the operation was chosen for three main reasons: (1) high water quality, (2) no native Atlantic salmon stocks present in the lake, and (3) positive support from the local surrounding community. The operation is made up of four separate cage blocks [(17,640 m^3 volumes), 8–12 cages/block (972–1452 m^2)] located 200 m from the shore in 20 m of water. The cages are made of 6-mm^2 mesh net and are also covered with a mesh net to prevent bird predation. This aquaculture location is licensed to hold 96 cages, not to exceed a total rearing volume of 28,224 m^3.

Each cage is stocked with approximately 1,500,000 disease-free certified Atlantic salmon parr (1- to 5-g average weight) with a maximum stocking density of approximately 4 kg/m^3. The parr are fed primarily during the day (to minimize feed waste), by hand or by automatic feeders. Current feed conversion rate (FCR) is approximately 1.15 from approximately 63 tons of feed over an 11-month period.

Every March approximately 1 million smolts (60-g average weight) are transferred from this site to the company's growout facility. The parr and smolts are transferred to and from the facility in trucks with carrying capacities of approximately 150,000 parr or 20,000 smolts each.

Public Health Issues Because this facility produces smolts for later growout at another facility, there are no direct human health issues associated with the operation. However, there are issues that should be considered even for this type of aquaculture operations to address possible antibiotic residues and chemical contaminants in the fish.

The growout farm is reducing the use of chemotherapeutants by using only disease-free certified high-health fish. Fish tissues are monitored for chemotherapeutant residue levels (ECS 2000). All chemotherapeutants are administered under the care of a veterinarian, and environmental monitoring programs are in place to analyze for these compounds (ECS 2000).

After antibiotic use, withdrawal periods of the fish range from 400 to 600 degree days (e.g., a degree day is the sum of temperatures for each day) (ECS 2000). In addition, parr require an additional 12 months of growout before they are ready for human consumption.

Animal Health Issues To minimize the likelihood of disease transfer to wild species, only certified disease-free fish are reared at the site. Each site is fallowed for 4 weeks before the sites are restocked. The fish are stocked at low densities to minimize stress and are monitored on a regular basis. Disease outbreaks are treated under the direction of a veterinarian. The aquacultured fish are vaccinated, and veterinarians regularly monitor their health condition. The aquaculture rearing site is monitored frequently by the Marine Institute of Fish Health to ensure compliance with their management plans (ECS 2000). To control diseases and protect the wild stock, dead fish are promptly removed from the cages, transported in leakproof containers, and disposed of in land-based rendering plants (ECS 2000).

Environmental Health Issues Escapements are a concern for environmental health, and contingency plans for the recapture of escaped salmon are in place. The primary species in the lake are eels and trout. No wild Atlantic salmon are present in the lake, although the potential for displacement of other native species by escaped Atlantic salmon is a possibility.

The farm uses chemotherapeutants such as formalin and malachite green to control parasites and fungus, and chloramine-T and sulfatrim for bacterial diseases. However, the farm is reducing its use of chemotherapeutants, and fish tissues are monitored for chemotherapeutant residue levels to ensure that they do not exceed the recommended levels. All chemotherapeutants are used under the care of a veterinarian, and monitoring programs are in place to analyze for these compounds in the environment (ECS 2000).

A comprehensive survey, before commencement of operation, was conducted on the watershed, sediments, and water quality of Lough Allen. The survey data included information on water and sediment quality from 1993 to 1999. These data are compared frequently with current water and sediment data collected from the farm, and they indicate that any effects from the farm operation are slight and are not of environmental concern (ECS 2000). Water quality monitoring is conducted to evaluate phosphorus loading to Lough Allen from the cages. Samples collected from the farm site and open water lake control sites indicate that except for occasional higher ammonia concentrations around the cages, no other significant differences in water quality parameters are present (ECS 2000). Future monitoring activities of this facility may include (1) monitoring of temperature and dissolved oxygen at the cage sites and at open water control sites; (2) analyzing water for transparency, chlorophyll, total phosphorus, soluble reactive phosphorus, and total oxidized nitrogen; (3) reporting and monitoring of weather conditions; and (4) survey of the sediment for total phosphorus and total nitrogen levels every 2 years (ECS 2000).

The cages, which are not visible from the shoreline, are monitored on a constant basis for any damage to their structure to minimize potential escapements by the cultured fish. In addition, when disease outbreaks occur, the company's management plan requires that only one cage be treated at a time with chemotherapeutants, to minimize impacts from any spillage of the chemotherapeutants (ECS 2000).

CONCLUSION

The future success of aquaculture in industrialized countries requires that current and new operations are developed and maintained in a sustainable and environmentally and socially friendly manner. Aquacultured products are proven to be safe and wholesome. Nevertheless, detailed planning efforts are essential to anticipate and manage potential public, animal, and environmental health issues associated with aquaculture products and the development and operation of aquaculture facilities. Central planning efforts must address these issues on a local, regional, national, and international basis.

REFERENCES

Abad, F.X., R.M. Pintó, R. Gajardo, and A. Bosch. 1997. Viruses in mussels: public health implications in depuration. J. Food Protection 60: 677–681.

Ahmed, F.E. 1991. Seafood Safety. Committee on evaluation of the safety of fishery products (ed. F.E. Ahmed). National Academy Press. Washington, DC.

Angot V. and Brasseur. 1993. European farms and Atlantic salmon (*Salmon salar L.*) are safe from anasakis larvae. Aquaculture 118: 339–344.

Alderman, D.J. and T.S. Hastings. 1998. Antibiotic use in aquaculture: Development of antibiotic resistance—potential for human health risks. Intl. J. Food Sci. Technol. 33: 139–155.

Anon. 2001. Marine farm finds ISA infected salmon. First Farming News Jan/Feb. 2001: 64.

Anon. 2000. Salmon monitoring program established. Fish Farming 13(2): 3.

Anon. 1995. OIE diagnostic manual for aquatic animal diseases. Office International des Epizooties, Paris.

Aoki, T. 1997. Resistance of plasmids and the risk of transfer. Furunculosis: multidisciplinary fish disease research (ed. E.M. Bernoth), pp. 433–440. Academic Press, London.

Aquaculture.com. 2000. Code of conduct for European aquaculture. Federation of European Aquaculture Producers. http://www.aquaculture.com.

Audicana, L., M.T. Audicana, L. Fernández de Corres, and M.W. Kennedy. 1997. Cooking and freezing may not protect against allergenic reactions to ingested *Anisakis simplex* antigens in humans. Veterinary Record: 140–235.

Baltic and Nordic News Service. 1999. Salmon lice are a greater threat to wild salmon in the Norwegian fjords than previously thought. In some fjords they kill 90% of the smelt, according to a new study. B&BNNS, Copenhagen, 11-08-99.

Bangen, M., K. Grave, R. Nordmo, and N. Soli. 1994. Description and evaluation of a new surveillance program for drug use in fish farming in Norway. Aquaculture 119(2/3): 109–118.

Bangen, M., K. Grave, and T. Horsberg. 1996. Surveillance of drug proscribing for farmed fish in Norway. Possible applications of computerized prescription information. J. Vet. Pharmacol. Therapeut. 19(1): 78–81.

Bernardet, J.F., B. Kerouault, and C. Michel. 1994. Fish Pathol. 29: 105–111.

Bisharat, N. and R. Raz. 1996. *Vibrio* infection in Israel due to changes in fish marketing. Lancet 348: 1585–1586.

Blake, P.A., M.H. Merson, R.E. Weaver, D.G. Hollis, and P.C. Heublein. 1997. Disease caused by a marine *Vibrio*: Clinical characteristics and epidemiology. N. Engl. J. Med. 300: 1–5.

Boesch, D.F., D.M. Anderson, R.A. Horner, S.E. Shumway, P.A. Tester, and T.E. Whitledge. 1997. Harmful algal blooms in coastal waters: Options for prevention, control and mitigation. NOAA Coastal Ocean Program. Decision Analysis Series No. 10. NOAA Coastal Ocean Office. 1315 East West Highway, Silver Spring, MD 20910. p. 50.

Boyd, C.E. 2001. Role of BMP in environmental management of aquaculture. Presented at the World Aquaculture Society. Lake Buena Vista, Florida, Jan. 21–25, 2001. Book of Abstracts: 73.

Boyd, E.E. 2000. U.S. Environmental protection agency meets on aquaculture effluents. Global Aquaculture Advocate June 2000, 3(3): 4.

Brady, Y.J. 2000. Viral diseases of fish and shellfish. *In*: Encyclopedia of Aquaculture (ed. R.R. Stickney), pp. 949–957. John Wiley & Sons, Inc., New York.

Cann, D.C., L.Y. Taylor, and G. Holls. 1975. The incidence of *Clostridium botulinum* in farmed trout in Great Britain. J. Appl. Bacteriol. 39: 331–336.

Centers for Disease Control and Prevention (CDC). 2000. Surveillance for Foodborne Disease Outbreaks—United States, 1993–1997. MMWR Morb. Mort. Wkly Rep. March 17, 2000/49 (SS01): 1–51.

Centers for Disease Control and Prevention (CDC). 1996. Invasive Infection with *Streptococcus iniae*—Ontario, 1995–1996. MMWR Morb. Mort. Wkly Rep. 1996; 45(30): 650–653.

Cole, M.T., M.B. Kilgen, L.A. Reily, and C.R Hackney. 1986. Detection of enteroviruses and bacterial indicators and pathogens on Louisiana oysters and their waters. Micro. Rev. 32: 61–79.

D'Aoust, J.Y. 1994. Salmonella and the international food trade. Intl. J. Food Microbiol. 24(1/2): 11–31.

Deardorff, T.L. and M.L. Kent. 1989. Prevalence of larval *Anisakis simplex* in pen reared and wild caught salmon (salmonidae) from Puget Sound. J. Wildlife Dis. 25: 416–419.

DeVoe, M.R. 1997. Marine aquaculture regulations in the United States: Environmental policy and management issues. *In*: Proc. Twenty-fourth U.S.-Japan Aquaculture Panel Symposium (ed. B. Jane Keller), pp. 1–15.

Corpus Christi, TX, Oct. 8–10, 1995, Texas A&M Univ., Sea Grant College Program, 1997. UJNR Technical Report, (24).

Ecological Consultancy Services Ltd. (ECS). 2000. Environmental impact statement for retention of existing salmon smolt farm on Lough Allen, County Leitrim. Ecological Consultancy Services Ltd. 17 Rathfarnham Road, Terenure, Dublin 6W, Ireland. www.ecoserve.ie/download/.

Elliot, E.L. and R.R. Colwell. 1985. Indicator organisms for estuarine and marine waters. FEMS Micro. Rev. 32: 61–79.

EMEA. 2000. www.eudra.org/en-home.htm. Committee for veterinary medicinal products, summary reports. European Agency for the Evaluation of Medicinal Products. London, UK.

Emerson, C. 1999. Aquaculture impacts on the environment. Hot topic series. Cambridge Scientific Abstracts. December.

Ervik, A., B. Tehorsen, V. Eriksen, B.T. Lunestad, and O.B. Samuelsen. 1994. Impact of administering antibacterial agents on wild fish and blue mussels (*Mytilus edulis)* in the vicinity of fish farms. Dis. Aquatic Organisms 18: 45–51.

FAO. 2000. The state of the world fisheries and aquaculture. ISBN 92-5-104492-9. Food and Agriculture Organization of the United Nations. Rome, Italy.

FAO. 1997. Review of the state of world aquaculture. FAO Fisheries Circular No. 886 FIRI/C886(Rev.1). ISSN 0429-9329. Food and Agriculture Organization of the United Nations. Fisheries Department, Rome, Italy.

FAO. 1995. Code of conduct for responsible fisheries. Food and Agriculture Organization of the United Nations, Fisheries Department, Rome, Italy. p. 41.

Federal Register Notice. 2000. PDF announcing proposed information collection request for the aquatic animal production industry. Sept. 14, 2000. Vol. 65, No. 179.

Fielder, H., K. Cooper, S. Bergek, M. Hjelt, C. Rappe, M. Bonner, F. Howell, K. Willett, and S. Safe. 1998. PCDD, PCDF, and PCB in farm-raised catfish from southeast United States—concentrations, sources and CYP1A induction. Chemosphere 37: 1645–1656.

Folke, C., N. Kaltsky, H. Berg, A. Jansson, and M. Troell. 1998. The ecological footprint concept for sustainable seafood production: A review. Ecological Applications. Special Supplement: 563–571.

Food Chemical News. 2001. Report to show only trace amounts of metals in farm raised fish. 42(51): 26.

Food and Drug Administration (FDA). 1998. Fish and fisheries products hazards and control guide. 3rd ed. FDA. Office of Seafood, Washington, DC.

Garrett, E.S., M.L. Jahncke, and R.E. Martin. 2000. Application of HACCP principles to address food safety and other issues in aquaculture: An overview. J. Aquatic Food Product Technol. 9(1): 5–20.

Garrett, E.S., M.L. Jahncke, and J. Tennyson. 1998. Microbial hazards and emerging food-safety issues associated with seafoods. J. Food Protection 60(11): 1409–1415.

Garrett, E.S., C. dos Santos, and M.L. Jahncke. 1997. Public, animal and environmental health implications of aquaculture. J. Emerging Infect. Dis. 3(4): 453–457.

Golz, W., R.F. Malone, and S. Chen. 1997. Reducing the environmental impact of high density fish production: An integrated approach to solids treatment for recirculating aquaculture systems using expandable granular biofilters. *In*: Proc. Twenty-fourth U.S.-Japan Aquaculture Panel Symposium (ed. B. Jane Keller), pp. 157–164. Corpus Christi, TX, Oct. 8–10, 1995, Texas A&M Univ., Sea Grant College Program, 1997. UJNR Technical Report, (24).

Hackney, C.R. and A. Dicharry. 1988. Seafood-borne bacterial pathogens of marine origin. Food Technol. 42(3): 104–109.

Hankins, J.A. 2000. Perspective on the role of government, industry, and research in advancing the environmental compatibility and sustainability of aquaculture. *In*: Proc. of the Third International Conference on Recirculating Aquaculture (eds. G. Libey, M. Timmons, G. Flick, and T. Rakestraw), pp. 80–87. Virginia Polytechnic Institute and State University, Roanoke, VA, July 20–23, 2000.

Hatt, J. 1998a. New strategies minimize fish farm impacts. Northern Aquaculture. Environment '98. pp. 10–11.

Hatt. J. 1998b. International standards help fish farms improve environmental management. Northern Aquaculture. Environment '98. pp. 19–20.

Hawke, J.P. 2000. Bacterial disease agents. *In*: Encyclopedia of Aquaculture. (ed. R.R. Stickney). John Wiley & Sons, Inc., New York, pp. 69–97.

Hayashi, R., K. Harada, S. Mitsuhashi, and M. Inoue. 1982. Conjugation of drug resistant plasmids from *Vibrio anguillarum* to *Vibrio parahaemolyticus.* Microbiol. Immunol. 26: 479–485.

Howgate, P. 1998. Review of public health safety of products from aquaculture. Intl. J. Food Sci. Technol. 33: 99–125.

Howgate, P. 1997. Review of the public health safety of products from aquaculture. A report prepared for the Fish Utilization and Marketing Service, Fisheries Utilization Division, FAO. Presented at the Joint FAO/WHO Study Group on Food Safety Issues Associated with Products from Aquaculture. Bangkok, Thailand. June, 1997.

Huss, H.H., A. Pederson, and D.C. Cann. 1974. The incidence of *Clostridium botulinum* in Danish trout farms. I. Distribution in fish and their environment. II. Measures to reduce contamination in fish. J. Food Technol. 9: 455–458.

ICES. 1999. Statistical analysis of fish disease prevalence data extracted from the ICES Environmental Data Center. *In*: Report of the ICES Advisory Committee on the Marine Environment, 1998, ICES Cooperative Report, No. 233.

Inglis, T. 2000. Scottish Environmental Protection Agency: Policy update on regulations and expansion of caged fish farming of salmon in Scotland. Tom Inglis, Acting North Region Director, Aug. 2000, SEPA 59/00.

Jahncke, M.L. and M.H. Schwarz. 2000. Application of Hazard Analysis and Critical Control Point (HACCP) principles as a risk management approach for recirculating aquaculture systems (RAS). *In*: Proc. of the Third International Conference on Recirculating Aquaculture (eds. G. Libey, M. Timmons, G. Flick, and T. Rakestraw), pp. 45–49. Virginia Polytechnic Institute and State University, Roanoke, VA, July 20–23, 2000.

Jahncke, M.L. and M.H. Schwarz. 1998. Employee, animal and public health issues associated with recirculating aquaculture systems. *In*: Proc. of the First International Conference on Food Safety, Albuquerque, NM, Nov. 16–18, 1998, pp. 207–213.

Jensen, G.L. and K.J. Greenless. 1997. Public health issues in aquaculture. OIE scientific and Technical Review. 16(2): 641–651.

Joint Subcommittee on Aquaculture (JSA). 1994. Guide to drug vaccine and pesticide use in aquaculture. Publication No. B-5085. Texas Agricultural Extension Service. Texas A&M University. p. 68.

Kerry, J., M. Hiney, R. Coyne, D. Cazabon, S. NicGabhainn, and P. Smith. 1994. Frequency and distribution of resistance to oxytetracycline in microorganisms isolated from marine fish farm sediments following therapeutic use of oxytetracycline. Aquaculture 123: 43–54.

Kilgen, M.B., M.T. Cole, and C.R. Hackney. 1988. Shellfish sanitation studies in Louisiana. J. Shellfish Res. 7: 527–530.

King, R.K. and G. Flick. 2000. The potential for the presence of bacterial pathogens in biofilm of recirculating aquaculture systems. *In*: Proc. of the Third International Conference on Recirculating Aquaculture (eds G. Libey, M. Timmons, G. Flick, and T. Rakestraw). pp. 335–347. Virginia Polytechnic Institute and State University, Roanoke, VA, July 20–23.

Leung, C.K. Y.W. Huang, and O.C. Pansorbo. 1992. Bacterial pathogens and indicators in catfish and pond environments. J. Food Protection 55: 424–427.

Lightner, O.V. 1999. The penaeid shrimp viruses TSV, IHHNV, WSSV and YHV: Current status in the Americas, available diagnostic methods and management strategies. J. Appl. Aquaculture 9: 27–52.

MacKinnon, B.M. 1997. Sea lice: A review. World Aquaculture 28(3): 5–10.

McKinney, L. 1997. The management of shrimp fisheries and mariculture—Balancing the risk of disease—Why everyone hates you and thinks that you are an idiot. *In*: Proc. of the 22nd Annual Seafood Sciences Tropical and Subtropical Technol. Soc. of the Am., Biloxi, MS, Oct. 5–7.

Monro, A.S.S. and I.F. Waddell. 1984. Furunculosis: Experience of its control in the sea water cage culture of Atlantic salmon in Scotland. ICES. CM/F:32.

Morikawa, T. 2000. Sustainable marine aquaculture and stock enhancement in Japan. Special summer of 1999 aquaculture workshop report. A workshop held in the NOAA Science Center Auditorium, Aug. 11–13, 1999. U.S. Dept. of Commerce.

Myron, R. 2000. The availability and use of chemotherapeutic sealice products. Cont. to Zoology. 69(1/2):109–118.

Nakajima, K. 1997. Viral diseases in marine aquaculture in Japan. *In*: Proc. Twenty-fourth U.S.-Japan Aquaculture Panel Symposium (ed. B. Jane Keller), pp. 139–143. Corpus Christi, TX, Oct. 8–10, 1995, Texas A&M Univ., Sea Grant College Program, 1997. UJNR Technical Report, (24).

Nakajima, T., M. Suzuki, K. Harada, M. Inoue, and S. Mitsuhashi. 1983. Transmission of R plasmids in *Vibrio anguillarum* to *Vibrio cholerae*. Microbiol. Immunol. 27: 195–198.

National Marine Fisheries Service (NMFS). 2000a. Fisheries of the U.S. Current Fisheries Statistics. 1999. U.S. Department of Commerce, NOAA, NMFS. Silver Spring, MD.

NMFS. 2000b. Wild Atlantic salmon in Maine protected as endangered species. News release. Nov. 13, 2000. www.nero.nmfs.gov/atsalmon/.

National Marine Pathogen Workshop. 2000. Unpublished report: Workshop for developing a National Marine Pathogen Plan. National Marine Fisheries Service, Pascagoula, MS.

Nicholson, B.L. 2000. Fish diseases in aquaculture. A White Paper. Dept. of Biochemistry, Microbiology and Molecular Biology, Univ. of Maine, Orono, ME 04469.

Paille, D., C. Hackney, L. Reily, M. Cole, and M. Kilgen. 1986. Seasonal variation in the fecal coliform population of Louisiana oysters and its relationship to microbiological quality. J. Food Protection 50(7): 545–549.

Pazos, F., Y. Santos, A.R. Macias, S. Nunez, and A.E. Toranzo. 1996. Evaluation of media for the successful culture of *Flexibacter maritimus*. J. Fish Dis. 19: 193–197.

Pike, A.W. and S.L. Wadsworth. 1999. Sea lice on salmonids, their biology and control. Adv. Parasitol. 44: 233–337

Plumb, J.A. 1999. Health maintenance and principal microbial diseases of culture fish. Iowa State University Press, Ames, IA.

Rappe, C., S. Bergek, H. Fielder, and K.R. Cooper. 1998. PCDD and PCDF contamination in catfish feed from Arkansas, USA. Chemosphere 36: 2705–2720.

Raynor, B. 2000. ASF press release upsets NB growers. Fish Farming May, 2000. 12(13): 1–2.

Richards, G.P. 1988. Microbiological purification of shellfish: A review of depuration and relaying. J. Food Protection 51: 218–251.

Rose, J.B. and M.D. Sobsey. 1993. Quantitative risk assessment for viral contamination of shellfish and coastal waters. J. Food Protection 56: 1043–1050.

Samocha, T.M. and A.L. Lawrence. 1997. Shrimp farms' effluent waters, environmental impact and potential treatment methods. *In*: Proc. Twenty-fourth U.S.-Japan Aquaculture Panel Symposium (ed. B. Jane Keller), pp. 33–57. Corpus Christi, TX, Oct. 8–10, 1995, Texas A&M Univ., Sea Grant College Program, 1997. UJNR Technical Report, (24).

Sandaa, R.A., V.L. Torsvik, and J. Grokooyr. 1992. Transferable drug resistance in bacteria from fish farm sediments. Can. J. Microbiol. 38: 1061–1065.

Sinnott, R. 2000. Cost of sea lice to Scottish salmon farmers. Technical manager, Trout Aquaculture.

Skladany, M. and C. Bailey. 1994. Aquaculture contributions to community development in the United States. Report prepared for the Office of Technology Assessment, United States Congress. A contribution to the OTA project: "Aquaculture: Food and renewable resources from U.S. waters."

Smith, P., M. Hiney, and O.B. Samuelson. 1994. Bacterial resistance to antimicrobial agents used in fish farming: A critical review of method and meaning. Annu. Rev. Fish Dis. 4: 273–313.

Son, R., G. Rusul, A.M. Sahilah, A. Zainuri, A.R. Raha, and I. Salrnah. 1997. Antibiotic resistance and plasmid profile of *Aeromonas hydrophilla* isolates from cultured fish. Tilapia (*Tilapia mossambica*). Lett. Appl. Microbiol. 24: 479–482.

Stewart, G.J. and C.D. Sinigalliano. 1990. Detection of horizontal gene transfer by natural transformation in native and introduced species of bacteria in marine and synthetic sediments. Appl. Environ. Microbiol. 56: 1818–1824.

Stickney, R.R. 2000. Encyclopedia of Aquaculture (ed. R.R. Stickney). John Wiley & Sons, Inc., New York, p. 1063.

Tamplin, M.L. 1997. Pathogens affecting consumer health and fishery resources. Presentation to the National Ocean Service. Washington, DC. March 19–20, 1997.

Thompson, A.G. 1990. The danger of exotic species. World Aquaculture 21 (1990): 25–32.

Tucker, C.S. 2000. Channel catfish culture. *In*: Encyclopedia of Aquaculture (ed. R.R. Stickney). pp. 153–170. John Wiley and Sons, Inc., New York.

United States Department of Agriculture (USDA). 1989. A margin of safety: the HACCP approach to food safety education. Project report. Information on legislative affairs. FSIS, Washington, DC.

Vaughn, J.M. and E.F. Landry. 1984. Public health considerations associated with molluscan aquaculture system: human viruses. Aquaculture. 39: 299–315.

Watanabe, T.T., Y. Aoki, Y. Ogata, and S. Egusa. 1977. R factors related to fish culturing. Ann. NY Acad. Sci. 182: 383–410.

Weinstein, M.R., M. Litt, D.A. Kertesz, P. Wyper, D. Ross, M. Coulter, A. McGreer, R. Facklam, C. Ostach, B.M. Willey, A. Borczyk and D.E. Low.

1997. Invasive infection due to a fish pathogen: *Streptococcus iniae*. N. Engl. J. Med. 33(7): 5589–594.

Weston, D.P. 1991. Aquaculture and water quality. The Effect of Aquaculture on Indigenous Biota. Baton Rouge, LA: World Aquaculture Society.

WHO. 1999. Food safety issues associated with products from aquaculture. Report of a Joint FAO/NACA/WHO Study Group. WHO Technical Report Series 883. WHO, Geneva, 1999, p. 55.

WHO. 1995. Control of foodborne trematode infections. WHO Technical Report Series 849. World Health Organization, Geneva.

Wolfe, J.C. 1998. Potential zoonotic infections in cultured food fish. *In*: Proc. of the 2nd International Conference on Recirculating Aquaculture (eds. G. Libey and M. Timmons), pp. 162–170. Virginia Polytechnic Institute and State University, Roanoke, VA, July 16–19.

Working Group on Marine Fish Culture (WGMAFC). 2000. ICES. Report of the working group on marine fish culture. ICES, CM. 2000/F:04.

Wyatt, L.E., R. Nickelson, and C. Vanderzant. 1979a. Occurrence and control of Salmonella in freshwater catfish. J. Food Sci. 44: 1067–1069,1073.

Wyatt, L.E., R. Nickelson, and C. Vanderzant. 1979b. Edwardsella trade in fresh cater catfish and their environment. Appl. Environ. Microbiol. 38: 710–714.

4

Hazard Analysis Critical Control Point and Aquaculture

Carlos A. Lima dos Santos

INTRODUCTION

The HACCP concept has taken on a genuinely global perspective in both the production and inspection of fish and fishery products. The system is to be applied throughout the food chain from primary production to final consumption, with its implementation being focused on foodborne risk demonstrated by sound science. However, in the case of the aquaculture sector, HACCP is being implemented at a more reduced pace globally, with the exception of a group of a few countries [including among others, the United States, Norway, New Zealand, Canada, Ireland, United Kingdom (Scotland), Cuba, Chile, Ecuador, Australia, Brazil, Indonesia, and Thailand] and some highly valued species (e.g., salmon, shrimp, trout).

Aquaculture is one of the fastest-growing food production systems in the world, resulting in an increased production of approximately 11% per year during the past decade. The increasing global importance

Public, Animal, and Environmental Aquaculture Health Issues,
Edited by Michael L. Jahncke, E. Spencer Garrett, Alan Reilly,
Roy E. Martin, and Emille Cole.
ISBN 0-471-38772-X (cloth)

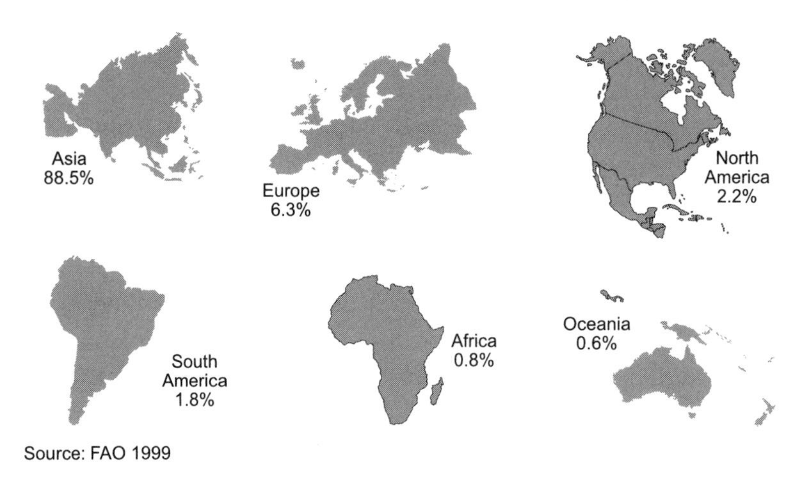

Figure 4.1. *Percent of Total Worldwide Aquaculture Production by Weight*

of aquaculture is directly related to its contribution in reducing the gap between supply and demand of fish and fishery products. Commercial aquaculture enterprises contribute significantly to the economies of many countries, where high-value species are a major source of foreign trade. Asia, which has the longest tradition in aquaculture, produces more than 90% of total aquaculture production by weight (82.3% by value), followed by Europe, North America, South America, the former USSR region, Africa, and Oceania (Fig. 4.1). In terms of economic country grouping, approximately 87.1% of the total aquaculture production was produced within developing countries and 76.8% from low-income food-deficit countries (LIFDC) where the per capita income was below US$1465 in 1995 (WHO 1999).

AQUACULTURE FOOD SAFETY AND QUALITY ISSUES

The aquaculture industry has achieved an excellent record regarding the production of safe, good-quality fishery products. The exceptional consistency of product quality and safety has been one of the factors that has made aquaculture so attractive.

FOOD SAFETY ISSUES

A variety of human health hazards may be associated with the consumption of wild-caught fish and their products. These concerns generally relate to biological and chemical hazards. Similar categories of hazards can be present in farmed fish. In addition, farmed fish may also have hazards relating to the presence of residues of veterinary drugs used for the treatment of fish diseases. Food safety issues associated with aquaculture products will differ not only from region to region but from habitat to habitat and will vary according to the production methods applied, management practices, environmental conditions, and cultural habits of food preparation and consumption.

Many of the issues related to aquacultured fishery product safety were considered by a Joint FAO/NACA/WHO Study Group, which met in Bangkok, Thailand, in July 1997. The Study Group considered food safety issues associated with farmed finfish and crustaceans, particularly biological and chemical hazards that may occur during the production of these aquatic products. The Study Group took into consideration the identification and quantification of hazards and implementation of control measures for potential safety hazards, including those of current national and international programs. Foodborne parasitic infections, foodborne diseases associated with pathogenic bacteria, residues of agrochemicals, veterinary drugs, and persistent polychlorinated hydrocarbons and heavy metal contamination were all identified by the Study Group as hazards of aquacultured products (WHO 1999).

As the aquaculture industry assumes an expanding role in meeting consumers' demands for fishery products, it will come under increased scrutiny by national agencies and international organizations whose responsibilities are to identify potential public health problems in food production and processing and, to the extent possible, to minimize their impacts through their food safety regulations. Thus, as aquaculture makes its transition to a major food-producing sector, proper assessment and control of any food safety concerns are becoming increasingly important (Reilly and Käferstein 1997; Howgate 1998; Reilly et al. 1997).

FOOD QUALITY ISSUES

Like traditionally caught fish, aquacultured products are perishable and require the same basic food handling and processing schemes to prepare them for entry into the marketplace. However, despite these similarities, aquaculturists want to separate their practices and products from the traditional fish and fishery product scenarios and, at times, a poor public image, for both esoteric and enhanced marketing purposes. This distinction is based on the premise that aquaculture can provide more controlled production resulting in better product quality. Because production, processing, and distribution of cultured products are under greater control than those of captured items, the opportunity for increased control of quality, particularly flavor quality, is now possible (Howgate 1997).

Aquaculture brings a variety of high-quality fishery products to the marketplace throughout the year. To many, the success of the industry rests on the consumer's perception of product quality. The advantages of aquaculture-raised fish include the production of safe and better-quality fishery products, which are being emphasized in consumer education programs in many countries. Aquaculturists anticipate increasing prices within the industry because of expanded demands for their products, which should provide favorable profit margins to compensate for previously prohibitive production costs. Although encouraged by these trends, many in the aquaculture industry did not anticipate the degree of product scrutiny that accompanies increasing demands. In North America consumers seem willing to pay higher prices for these products, and they expect better product quality and safety (Otwell 1989; NMFS 1991).

THE HACCP CONCEPT

Traditionally, the means of controlling risks of foodborne illness at the production level has been accomplished by inspection and surveillance of final products and by concentrating food safety efforts on end-product testing. This approach has clearly been a tactical error, because even the most careful and thorough inspection program and final product testing schemes will never lead to the proper management of risks (Huss 1994, 1998). According to Mossel et al. (1997), it is almost inconceivable that the inspection and testing approach has endured for more than 80 years.

The fallacy of relying on end-product testing has also been addressed by Rainosek (1997), who pointed out that when sampling, there always exists the possibility of making an incorrect decision on the actual conforming status using a predetermined standard. In this context, Rainosek further explained that inadequate understanding of the limitation of a sampling plan may lead to a false sense of security by those responsible for its use. For example, when a company or inspection agency employs the commonly used sampling plan, $n = 6$, $c = 1$ (where n is the sample size and c is the acceptance number, i.e., accept the entire lot if none or 1 of the 6 sampled items are nonconforming), there is a

TABLE 4.1. The Probability for Lot Acceptance, Illustrating the "10 Percent Rule" Regarding Lot Size

N	P_a
100	84.90%
1,000	84.48%
10,000	84.44%
100,000	84.44%

50% chance of accepting an entire lot that has 26.4% nonconforming items.

Rainosek also addressed the question, "What effect does lot size have on the performance characteristics of sampling plans?", the answer to which can be unbelievable to those not statistically informed, pointing out that when the sample size is less than 10% of the lot size, the probability of lot acceptance is negligibly affected by lot size. For example, suppose a lot contains 12% nonconforming sample units (items) and the single sampling plan, $n = 6$, $c = 1$, is used. The probabilities of lot acceptance (P_a) for different values of lot size (N) are given in Table 4.1.

Further, Rainosek elaborates on the practice of resampling, that is, when the first sample indicates lot rejection, pull another sample and take action on the lot based only on the results of the second sample. The net effect of resampling (which is often employed by those not familiar with the unintended consequences) is to actually increase the chance of accepting noncomplying products. An example of this increased adverse effect and reaching an incorrect decision can be shown in the single sampling plan, $n = 5$, $c = 0$. In this instance, when 20% of the lot is nonconforming such a plan will indicate lot acceptance 33% of the time. However, when resampling is employed (pull a second sample of $n = 5$, $c = 0$ if the first sample indicates lot rejection), the chance of accepting a lot with 20% nonconforming units will increase from 33% to 55%. Rainosek further points out the need not to confuse "resampling" with the legitimate use of "double sampling plans," in which lot acceptance, rejection, and additional sampling are options based on the structure of the double sampling plan. Without a doubt, relying on sampling and testing of foods for compliance can be a risky enterprise. A more preventive approach for controlling hazards such as HACCP represents a significant advancement in consumer protection.

In contrast to the traditional food control approach, a study leading to the control of all factors related to every stage of the food chain comprises what is known as the HACCP approach (Huss 1994). The HACCP system, which is science based and systematic, not only identifies specific hazards but also measures for their control to ensure the safety of food. HACCP is a tool used to assess hazards and to establish control systems that focus on prevention rather than relying mainly on end product testing. HACCP can be applied through the food chain from primary production to final consumption, and its implementation should be guided by scientific evidence of risks to human health. In addition, the application of HACCP systems can aid control by regulatory authorities and promote international trade by increasing confidence in safety of traded foods (CAC 1997).

Garrett and Hudak-Roos (1992) stated that it must be understood that "HACCP is a nontraditional inspection system. It is a system that does not require continuous inspection, and as such, separates the nice from the necessary, or the essential from the nonessential. This separation allows proper focusing of limited resources. Under HACCP the inspectional frequency should be much less than that currently employed under the traditional inspectional approach, GMP, or relying on end product examination when the product may have been produced under unknown hygienic operations such as would be the case with imports."

HACCP was considered a superior method of fish inspection by the participants of the International Conference on Quality Assurance in the Fish Industry held in Lyngby, Denmark in 1991. Participants agreed that the HACCP concept should be applied in the fish industry to cover food safety, plant/food hygiene, and economic fraud issues (FAO 1992). During the Second International Conference on Fish Inspection and Quality Control held in Washington, DC in 1996, participants affirmed that HACCP-based programs were in the process of being implemented on a global scale. Governments and industry alike were urged by this international conference to continue their efforts and to give a high priority to the full implementation of HACCP-based systems.

It has been reported by FAO (2000) that a large number of countries now have specific HACCP-based regulations regarding the safety of fish and fish products, including products from aquaculture. Approximately 65% of the total international fish trade is performed under HACCP-based regulations. The large exception is the Japanese market. Japan, which accounts for about 32% of the total fish market (demand) has no HACCP regulations yet. Canada, the European Union (EU), and the United States were first in adopting the HACCP-based regulations of fish and fish products; however, the shifting to the HACCP-based system imposed by regulations is not an EU and US phenomenon. All developed countries and a large number of developing countries have already shifted to HACCP-based systems.

HACCP AND AQUACULTURE

Although many agriculture experts believe that the application of HACCP may be difficult at the "farm level," a number of authors have considered the application of HACCP in aquaculture to be suitable (Kim 1993; Garrett and Jahncke 1992; Reilly and Käferstein 1997; Jensen and Martin 1997; Suwanrangsi 1997; Boyette 1997; Tookwinas and Suwanrangsi 1997; ISMEA 1997; Libey 1998; Lima dos Santos and Tacon 1998; Smith 1998; Antonetti et al. 1999; Lima dos Santos 1999; Garrett et al. 2000; Jahncke and Schwarz 2000).

A historical background of HACCP related to aquaculture in the United States is given by Jensen and Martin (1997). According to these authors, the first exposure of the US aquaculture community to HACCP occurred during 1989 and 1990 when three Aquaculture Application Workshops were conducted by the NFI in cooperation with the USNMFS. The first HACCP Regulatory Model for Aquaculture was published and released by USNMFS (in 1991). In 1992 the Mississippi University Extension Service produced a video series that was based on the HACCP Regulatory Model.

According to the Joint FAO/NACA/WHO Study Group (WHO 1999), when a hazard analysis vs. end use of aquaculture products is generically applied one may conclude that with the exception of those products eaten raw, in particular molluscan shellfish and some parasitized freshwater fish species (e.g., Cyprinidae in southeast Asia), aquaculture presents no more, and perhaps even less, of a safety risk than wild-caught fish and that aquaculture products may present some unique risks such as antibiotic residues and therapeutic residuals.

A draft Code of Hygienic Practice for the Products of Aquaculture was prepared by FAO and is presently under review by the Codex Committee on Fish and Fishery Product. This draft code includes specific HACCP considerations that would be used in conjunction with the Recommended International Code of Practice General Principles of Food Hygiene and the Annex, Hazard Analysis and Critical Control Point (HACCP) System and Guidelines for its Application. It is anticipated that this document will give particular attention to biological agents such as bacteria (*Salmonella* spp., *Shigella* spp., *Vibrio* spp.) and parasites (*Clonorchis sinensis, Opisthorchis* spp.) of human health significance, chemical contaminants (heavy metals, agricultural pesticides, industrial chemicals), and residues of veterinary drugs (antibiotics, parasite control agents).

PRACTICAL EXPERIENCES OF HACCP APPLICATION IN AQUACULTURE

According to published information, HACCP systems are being put into practice in aquaculture at various levels but mainly in the sectors of high-valued farmed species such as (1) salmon in Norway, Canada, Ireland, the United States, New Zealand, the United Kingdom, and Chile; (2) shrimp in Thailand, Ecuador, Australia, Cuba, Brazil, Central American countries, and the United States; (3) trout in European countries, Argentina, Peru, and Brazil; and (4) catfish in the United States, crawfish in the United States, and bullfrogs in Brazil. The United States has the largest volume of information and guidelines on how to apply HACCP in aquaculture, with catfish farming receiving the major coverage (Kim 1993; Boyette 1997).

In the case of developing countries, the application of the HACCP concept in aquaculture is mainly influenced by the need to comply with the sanitary requirements of main importing countries. Serious restrictions imposed by Japan on the importation of farmed shrimp contaminated with residues of veterinary drugs, particularly antibiotics, forced government and producers/exporters in Thailand, Indonesia, and the Philippines to implement the new control systems. The same restrictions were also imposed by the US and the EU, accelerating the process, which is also influenced by the "never-changing" zero tolerance level imposed by these markets concerning the presence of *Salmonella* in raw shrimp. Another common effort is directed to control sodium bisulfite residues in frozen shrimp resulting from dipping shrimp in sodium bisulfite solutions at farm level in an attempt to minimize "black spot."

Producers and governments alike, in developing countries, were influenced by the developments in the United States and Norway in their initial attempts of implementing the HACCP system. It must be emphasized that the majority of these countries chose to apply the HACCP system using the USNMFS comprehensive consumer hazard approach, that is, to control safety, quality and economic integrity hazards, as opposed to applying HACCP only to address seafood safety concerns.

The HACCP concept was first introduced to developing countries through FAO training programs on fish technology and quality assurance. The three USFDA/USNMFS International Seminars on the application of HACCP in the fish industry, with the participation of FAO, which were held in Kuala Lumpur, Brussels, and Mexico City (during 1990 and 1991), generated awareness of the system.

The FAO began assisting developing countries in their application of HACCP in the field of aquaculture during 1991 and 1992. This work was initially concentrated in Latin America and was accomplished by close collaborations with the USNMFS. Salmon farmers in Chile and trout farmers in Argentina and Brazil also used the experiences gained from the quality assurance methods applied in Norway (Valset 1997).

In Brazil, the number of farms using HACCP has gradually increased, due in part to the support by the Brazilian government projects being advanced by that country's Federal Fish Inspection Service. Shrimp farmers are leading the implementation of HACCP and are being closely followed by trout, tilapia, bullfrog, and shellfish bivalve farmers. Numbers of HACCP regulated aquaculture operations are small but are steadily growing. Six shrimp farms, two trout farms, one tilapia farm, and four bullfrog farms have operational HACCP plans. The implementation of the HACCP system in shrimp farms has been accelerated by new farms being established in the northeast states of Ceará, Piauí, and Rio Grande do Norte, Brazil. Three national training courses were held specifically on the design and application of HACCP in the aquaculture sector. The main goal of these training courses was to develop generic HACCP plans for instructional purposes, to address and control chemical contaminants, food additives, veterinary drugs, pesticides, heavy metals, and pathogenic bacteria.

In Chile, the government developed a Guideline for the Control of Veterinary Drug Residues in Aquacultured Products [National Fisheries Service (SERNAPESCA) 2000] based on HACCP principles. This Guideline offers excellent reference material to those who may be interested in implementing the system.

The application of HACCP in aquaculture in Asia is mostly restricted to the Association of Southeast Asian Nations (ASEAN) countries (Brunei, Indonesia, Malaysia, Philippines, Singapore, Thailand, and Vietnam). These countries received substantial assistance from the ASEAN-Canada Fisheries Post-Harvest Technology Project, which worked together with the ASEAN-associated governments and industries of these countries in the design and implementation of the system (Lima dos Santos and Tacon 1998). The technical approach observed in the HACCP plans reflects the multiple influences from Canada (Quality Management Program—QMP), the United States (initially from the USNMFS approach and today from USFDA), Australia, New Zealand, and Norway.

Thai efforts in implementing the application of HACCP in shrimp aquaculture deserve special mention. The Department of Fisheries in Thailand, together with the aquaculture and shrimp processing industry, developed a National Quality Management Plan specifically designed to control drug and chemical residue contamination in farmed shrimp and to prevent microbiological contamination. The HACCP concept is being successfully introduced into all shrimp aquaculture operations, including production and handling at the farm level (Suwanrangsi 1997; Tookwinas and Suwanrangsi 1997).

In Africa, the implementation of HACCP in aquaculture is almost exclusively limited to bivalve shellfish growing areas of north African countries (Morocco and Tunisia) and are based on the European approach.

HACCP IN THE CONTROL OF HUMAN PATHOGENIC FISHBORNE PARASITES

A large number of both marine and freshwater fish species can serve as a source of foodborne parasitic infections. Some of these parasites are highly pathogenic, with the main cause of human infection being that of consumption of raw or undercooked fish. It is evident that these infections are only prevalent in a very few countries of the world, essentially among the communities where eating raw or inadequately cooked fish is a cultural habit. Fish are the intermediary hosts of these parasites, and humans, along with other mammals, are a definitive host of the parasite. The main human diseases caused by these parasites are trematodiasis, cestodiasis, and nematodiasis.

Fishborne trematodiasis (FBT) is a serious disease in various parts of the world, particularly in Asia. Although the disease is seldom fatal, trematodes can cause morbidity and serious complications among humans, leading to death in immunocompromised individuals (such as liver cancer in patients infected with *Opisthorchis viverrini*). The route of infection is through the ingestion of viable encysted metacercariae parasites, which are generally found in the flesh of raw, undercooked, or semiprocessed freshwater fish. The two major genera important to human health are *Clonorchis* and *Opisthorchis*.

During the WHO Study Group Meeting on the Control of Foodborne Trematode Infections held in Manila, Philippines, during October 1993, an attempt was made to design a new strategy for the prevention and control of FBT in cultured fish based on the HACCP plan for the production and marketing of fish free from FBT infection. The farmer would be technically assisted by the local public health officers, fisheries officers, and fish inspection officers in implementing such procedures (WHO 1995; Lima dos Santos 1994).

Experiments to control and/or prevent foodborne trematode infections, employing the HACCP concept, were carried out by FAO in Asia by a multidisciplinary team of experts in public health, parasitology, aquaculture, fisheries extension, and fish inspection. During studies in Thailand and Vietnam, experimental activities were conducted simultaneously in two side-by-side fish ponds. In the experimental ponds, fish were cultured according to HACCP principles, whereas fish in the control pond were cultured according to conventional, local aquaculture practices. Water supply, fish fry, fish feed, and pond conditions in the experimental pond were identified as critical control points (CCP). The HACCP principles of hazard analysis, preventive measures, critical limits, monitoring, record keeping, and verification procedures relating to the CCP were applied. Preliminary results showed that no fish infected with the parasite metacercariae were observed in the experimental ponds. These results indicate that application of HACCP-based principles to carp culture in both countries can be an effective way to control and/or prevent FBTs. Further studies are recommended to confirm these preliminary results (Kamboonruang et al. 1997; Son et al. 1997)

HACCP FOR THE CONTROL OF AQUATIC ANIMAL DISEASES

The development and sustainability of aquaculture are largely at stake as significant ecological and pathological problems are on the increase in the vast majority of producing countries. Prevention and control of diseases are now a priority for the durability of this industry (Bachère 2000).

Shrimp farming constitutes an important source of revenue and employment in many countries, particularly in developing countries. However, infectious diseases have affected the profitability of the industry. For this reason, disease prevention is a priority in the shrimp industry and has become a prime area of research (Rodriguez and Le Moullac 2000).

The causative agents of shrimp diseases are mainly viruses and/or bacteria belonging to the family of Vibrionaceae. These pathogens hamper larval production, which leads to profitability problems due to stock mortalities. The practice of shrimp transfers between water areas at both the national and international level has contributed to the spread of diseases that are now enzootic (Bachère 2000).

To a great extent, the control of shrimp diseases depends on establishing an equilibrium between (1) the environmental quality, (2) preventive control measures of diseases, and (3) the health status of the shrimp. The problems involved are complex, and their solution requires a multidisciplinary approach. In addition to food safety issues, HACCP principles provide a disciplined scientifically based structure that can be used to develop risk management programs addressing a variety of issues. Within this framework, HACCP has been considered as a possible tool for the control of shrimp diseases caused by viruses (Jahncke 1996; Lima dos Santos 1998; Jahncke et al. 2000).

The idea of applying HACCP to address aquatic animal diseases is revolutionary and is currently being practiced at experimental levels. In the United States, a preliminary risk management program using HACCP principles has been developed to control shrimp viruses at shrimp production and processing facilities. A HACCP risk management plan was developed for a vertically integrated shrimp culture operation consisting of breeding brood stock production, maturation/reproduction, larval culture, nursery, and growout capabilities. Outsourced shrimp, feed, and incoming water along with effluent water at pond growout within the plan were identified as CCP. Likewise, an aquatic animal disease prevention HACCP plan with appropriate models was developed for a shrimp processing facility that produces 4500 kg of shrimp per hour and uses 3.8×10^5 liters of water per hour. Shrimp receipt, solid protein recovery, transport for solids processing, and wastewater discharge were also identified as CCP in this HACCP plan (Jahncke et al. 2001).

A preliminary study was completed in Sri Lanka by FAO and the local ministry of fisheries concerning the possibility of applying HACCP principles to control "white spot" disease (white spot syndrome virus) in farmed shrimp. The preliminary indications for the study were positive, despite the difficulties involved in the application of HACCP systems during the study (Lima dos Santos 1998).

CONCLUSIONS AND RECOMMENDATIONS

Identification of ways and means to bring about the application of HACCP-based food safety and quality control systems, especially in subsistence types of aquaculture, is a major challenge. It may not be difficult to apply the HACCP-based system to large-scale commercial aquaculture ventures. However, application of HACCP-based systems in small-scale subsistence aquaculture systems where fish are mainly farmed for domestic consumption, under minimal technical inputs, knowledge, and assistance, will certainly pose considerable difficulties. Therefore, according to WHO (1999), the design and implementation of such systems should be considered only after a careful evaluation of the feasibility of applying such a control system to a particular aquaculture system and the risks associated with the system's components and procedures and after identification of the correct and appropriate CCP.

In developing countries HACCP application in aquaculture is primarily limited in practice to farming activities that are linked to international trade. Government and industry actions are geared by regulatory initiatives at national and international levels to problems related to consumer safety (residues of veterinary drugs and pesticides, new microbial concerns), product quality standards and product identification, aquatic animal health (control of the spread of fish diseases), and socioeconomic issues (aquaculture vs. environmental degradation, aquaculture vs. tourism, etc.). Although the application of HACCP principles at aquaculture sites is in its infancy, the concept has the potential to make an excellent contribution in the control of animal and human pathogens. Extending the HACCP concept to aquaculture sites presents an important challenge and will occur when major national and international efforts are sufficiently focused and concentrated to do so. It has been proven that application of the HACCP concept, at the processing level, is a superior approach over traditional methods to control food safety and quality hazards, but continued efforts in the training and understanding of HACCP concepts are still needed at the processing level to ensure the realization of the full benefits of the concept.

REFERENCES

Antonetti, P., G. Canetti, and M. Doimi. 1999. The HACCP system and aquaculture. "Il Pesce" ("The Fish") No. 2, pp. 45–49 (in Italian).

Bachère, E. 2000. Shrimp immunity and disease control. Aquaculture 191 (1–3): 3–11.

Boyette, K.D. 1997. Quality Control and HACCP in the catfish industry. *In*: Fish Inspection, Quality Control and HACCP: A Global Focus (eds.

R.E. Martin, R.L. Collette, and J.W. Slavin), pp. 388–391. Technomic Publishing Company, Inc., Lancaster, PA.

Bryan, F.L. 1992. Hazard Analysis Critical Control Point Evaluations: a guide to identifying hazards and assessing risks associated with food preparation and storage. World Health Organization, Geneva.

CAC. 1997. Hazard Analysis and Critical Control Point (HACCP) System and Guidelines for its Application. Codex Alimentarius Commission (CAC) Food Hygiene Basic Texts. p. 58. FAO/WHO, Rome.

FAO. 1992. Report of the FAO/DANIDA Expert Consultation on Quality Assurance in the Fish Industry, Lyngby, Denmark, 26 August–6 September 1991, FAO/Denmark Cooperative Programme. FAO Fisheries Report No. 479 (FIIU/R479), p. 37. Rome, Italy.

FAO. 2000. Commodity Market Review 1999–2000. Rome, Italy. p. 88.

Garrett, E.S. and M. Hudak-Roos. 1992. The US model seafood surveillance project. *In*: Quality Assurance in the Fish Industry. pp. 501–508. Elsevier Science Publishers BV, Amsterdam, The Netherlands.

Garrett, E.S. and M.L. Jahncke. 1994. HACCP in Aquaculture. Paper presented at the IAMFES Meeting, San Antonio, TX, August 3, 1994.

Garrett, E.S., C.A. Lima dos Santos, and M.L. Jahncke. 1997. Public, Animal, and Environment Health Implications of Aquaculture. Emerg. Infect. Dis. 3(4): 453–457. October–December 1997.

Garrett, E.S., M.L. Jahncke, and R. Martin. 2000. Applications of HACCP Principles to Address Food Safety and Other Issues in Aquaculture: An Overview. J. Aquatic Food Product Technol. 9(1): 5–20.

Howgate, P.C. 1997. Review of the Hazards and Quality of Products from Aquaculture. A report prepared for the Fish Utilization and Marketing Service, Fisheries Industry Division, FAO of the United Nations, Rome, Italy, p. 48.

Howgate, P.C., C.A. Lima dos Santos, and Z.K. Shehadeh. 1997. Safety of Food Products from Aquaculture. *In*: Review of the State of World Aquaculture. FAO Fisheries Circular No. 886, Rev. 1, FAO of the United Nations, Rome, Italy. p. 163.

Huss, H.H. 1994. Assurance of Seafood Quality. FAO Fisheries Technical Paper No. 334. FAO of the United Nations, Rome, Italy. p. 169.

ISMEA. 1997. Feasibility study on the possibility of introducing quality systems in the aquaculture sector. ISMEA (Institute for the Study, Research, and Information on Agricultural Marketing), Rome, Italy. p. 80.

Jahncke, M.L. 1996. The application of the HACCP concept to control exotic shrimp viruses. *In*: Proc. of the NMFS Workshop on Exotic Viruses. New Orleans, LA, June, 1996.

Jahncke, M.L., C.L. Browdy, M.H. Schwarz, A. Segars, J.L. Silva, D. Smith, and A. Stokes. 2001. Preliminary application of hazard analysis critical control point (HACCP) principles as a risk management tool to control exotic viruses at shrimp production and processing facilities. *In*: The New Wave, Proc. of the Special Session on Sustainable Shrimp Culture, Aquaculture 2001 (eds. C.L. Browdy and D.E. Dory), pp. 279–284. The World Aquaculture Society, Baton Rouge, LA.

Jahncke, M.L. and M.H. Schwarz. 2000. Application of Hazard Analysis and Critical Control Point (HACCP) principles as a risk management approach for recirculating aquaculture systems (RAS). *In*: Proc. of the Third International Conference on Recirculating Aquaculture (eds. G. Libey, M. Timmons, G. Flick, and T. Rakestraw), pp. 45–49. Virginia Polytechnic Institute and State University, Roanoke, VA.

Jensen, G.L. and R.E. Martin. 1997. The impacts of federal Food Safety Initiatives on the aquaculture producer-processor HACCP juncture. *In*: Proc. of the Twenty-First Annual Meeting of the Seafood Science and Technology Society of the Americas (held jointly with The Atlantic States Fisheries Technology Society). Florida Sea Grant College Program. Gainesville, FL.

Khamboonruang, C., R. Keawivichit, K. Wongworapat, S. Suwanrangi, M. Hongpromyart, K. Sukhawat, K. Tonguthai, and C.A. Santos dos Lima. 1997. Application of Hazard Analysis and Critical Control Point (HACCP) as a possible control measure for *Opisthorchis viverrini* infection in cultured carp (*Puntius goniotus*). Southeast Asian J. Trop. Med. Publ. Hlth. 28 (suppl. 1), 1997, pp. 1164–1167.

Kim, J. 1993. Application of HACCP System to catfish producers and processors. *In*: Proc. of the Aquaculture Products Safety Form. HACCP Digital Library, National Sea Grant Depositary MASGC-W-93-001 pp. 125–134. (http://www.nsgd.gso.uri.edu/cgibin/copyright?/source/masgcw93001p125-134.pdf)

Libey, G.S. 1998. Development and implementing a HACCP Program for recirculating aquaculture systems: Production Issues. *In*: Proc. from the Success and Failures in Commercial Recirculating Aquaculture Conference (eds. G. Libey and M. Timmons), pp. 329–343. Northeast Regional Agricultural Engineering Service (NRAES), Cooperative Extension, Ithaca, NY.

Lima dos Santos, C.A. 1994. The possible use of HACCP in the prevention and control of foodborne trematode infections in aquacultured fish. Presented at the Symposium on New Developments in Seafood Science and Technology, 37th Annual Conference of the Canadian Institute of Food Science and Technology, Vancouver, Canada, May 15–18.

Lima dos Santos, C.A. 1995. Quality norms for Aquaculture products: trends on restriction problems. *In*: Marketing of aquaculture products. *In*: Proc. of the Seminar of the International Centre for Advanced Mediterranean Agronomic Studies CIHEAM Network on Socio-Economic and Legal Aspects of Aquaculture in the Mediterranean, Thessaloniki, Greece October 1995. Cashiers Options Mediterranéennes Volume. CIHEAN Zaragoza, Spain, pp. 85–92.

Lima dos Santos, C.A. and A. Tacon. 1998. Global trends responses of non-industrialized aquaculture producing countries to HACCP concepts and requirements. Paper presented at the World Congress of Aquaculture, Las Vegas, NV, February 15–19.

Lima dos Santos, C.A. 1999. HACCP and aquaculture: application in developing countries. Paper prepared at the II Venezuelan Congress of Food Science and Technology, Caracas, Venezuela, 24–28 April 1999. (In Spanish).

Mossel, D.A., C.B. Struijk, and J.T. Jansen. 1997. A rationale for containment of food-transmitted diseases of microbial etiology. Food Australia 49(5): 231–235.

National Advisory Committee on Microbiological Criteria for Foods. 1998. Hazard Analysis and Critical Control Point principles and application guidelines. J. Food Protection 61: 1246–1259.

NMFS. 1991. HACCP Regulatory Model—Aquaculture. Model Surveillance Project. National Seafood Inspection Laboratory. National Marine Fisheries Service. Pascagoula, MS, p. 65.

Otwell, W.S. 1989. Regulatory status of aquacultured products. Food Technol. Nov. 1989: 103–105.

Rainosek, A.P. 1997. Inherent risks in acceptance sampling. *In*: Fish Inspection, Quality Control, and HACCP: A Global Focus (eds. R.E. Martin, R.L. Collette, and J.W. Slavin), pp. 530–537. Technomic Publishing Co., Inc., Lancaster, PA.

Reilly, A. and F. Käferstein. 1997. Food safety hazards and the application of the principles of the hazard analysis and critical control point (HACCP) system for their control in Aquaculture production. Aquaculture Res. 1997(28): 735–752.

Reilly, A., P.C. Howgate, and F. Käferstein. 1997. Safety hazards and the application of the Hazard Analysis Critical Control Point System (HACCP) in Aquaculture. *In*: Fish Inspection, Quality Control and HACCP: A Global Focus (eds. R.E. Martin, R.L. Collette, and J.W. Slavin), pp. 353–373. Technomic Publishing Company, Inc., Lancaster, PA.

Rodriguez, J. and G. De Moullac. 2000. State of the art of immunological tools and health control of penaeid shrimp. Aquaculture 191(1–3): 109–119.

SERNAPESCA. 2000. Guidelines for the Design of Quality Systems for Aquaculture. Aquaculture. SERNAPESCA (Fisheries National Service), Ministry of Economy, Santiago, Chile.

Smith, S.A. 1998. HACCP Program for disease and therapeutics for intensively cultured foodfish. *In*: Proc. from the Success and Failures in Commercial Recirculating Aquaculture Conference, Vol. 2. Roanoke, VA, 19–21 July 1996 (eds. G. Libey and M.B. Timmons), Northeast Regional Agricultural Engineering Service (NRAES), Cooperative Extension, Ithaca, NY. pp. 344–345.

Son et al. 1997. Application of Hazard Analysis Critical Control Point (HACCP) as a possible control measure against *Clonorchis sinensis* in cultured silver carp *Hypophthalmichthys molitrix*. Paper presented at the FAO/NACA/WHO Study Group on Public Health Aspects of Products of Aquaculture, Bangkok, Thailand, July 1997.

Suwanrangsi, S. 1997. Hazard Control for Aquacultured Shrimp Products. *In*: Quality Management for Aquacultured Shrimp. ASEAN-Canada Fisheries Post-Harvest Technology Project-Phase II. Marine Fisheries Research Department, Singapore. pp. 1–26.

Tookwinas, S. and S. Suwanrangsi. 1997. Hazard control for aquacultured shrimp products. *In*: Fish Inspection, Quality Control and HACCP: A Global Focus (eds. R.E. Martin, R.L. Collette, and J.W. Slavin), pp. 380–387. Technomic Publishing Company, Inc., Lancaster, PA.

Valset, G. 1997. Norwegian hazard controls for aquaculture. *In*: Fish Inspection, Quality Control and HACCP: A Global Focus (eds. R.E. Martin, R.L. Collette, and J.W. Slavin), pp. 392–402. Technomic Publishing Company, Inc., Lancaster, PA.

WHO. 1995. Control of Foodborne Trematode Infections. Report of a WHO Study Group. WHO Technical Report Series No. 849, WHO, Geneva. p. 157.

WHO. 1999. Report of the Joint FAO/NACA/WHO Study Group on Food Safety Issues Associated with Products from Aquaculture, Bangkok, Thailand, 22–26 July 1997, WHO, Technical Report Series No. 883, WHO, Geneva.

5

Aquaculture and International Trade Regulations

E. Spencer Garrett

INTRODUCTION

Aquaculture and food standardization are not new technologies or endeavors, but rather have well-established roots that were conceived thousands of years ago. The evolution of aquaculture was recently described by Lovell (2000), beginning with fish farming being practiced in China as early as 2000 BC, and certainly by 475 BC, and during the first century AD, the Romans built fish ponds. Likewise, it is not much of a stretch to believe that the ancient Egyptian fishery methods reportedly described by Diodorun Siculus in Darby et al. (1997) led to the evolution of Mediterranean aquaculture during classical times.

Down through the ages there have been various religious concerns relative to food safety and ethical implications in the ingestion of foods, including seafoods, that have led to both Jewish (kosher) and Muslim (halal) laws that govern the processing, handling, sales, and marketing of foods (Regenstein et al. 2000) that can include aquaculture products.

Public, Animal, and Environmental Aquaculture Health Issues,
Edited by Michael L. Jahncke, E. Spencer Garrett, Alan Reilly,
Roy E. Martin, and Emille Cole.
ISBN 0-471-38772-X (cloth)

It has been historically recognized that as with any foodstuff, aquatic products (for which there are many divergent species) must have standardized criteria to determine their suitability for international trade relative to food safety, quality, and weights and measures.

Consumer protection in the context of fishery products can be grouped into three basic categories: (1) food safety, (2) food hygiene (harvested, prepared, packed, transported, and distributed in a sanitary or acceptable hygienic fashion), and (3) economic fraud (the misrepresentation of a food's quantity and/or its economic value) (Garrett et al. 1991). A historical progression of attempts by governments and regional organizations to promote consumer protection in the consumption of foods has been chronicled by the Codex Alimentarius International Foods Standards Programme (Codex). That narrative traces the ancient documentation of governmental authorities codifying rules to guard against dishonest practices in food sales. Methods to determine the correct weights and measures for food grains can be found in Assyrian tablets. Likewise, it has been reported that Egyptian scrolls prescribed labeling requirements for certain foods, the Athenian Greeks had standards and inspection approaches to control purity and soundness of beer and wines, and the Romans had well-organized consumer protection food control systems. Furthermore, in Europe, during the Middle Ages, individual countries passed laws concerning food safety and quality considerations for eggs, sausages, cheese, beer, wine, bread (FAO/WHO 1999a), and fish (Herborg 1997).

The international trade in fishery products (including aquaculture products) is massive, complex, and not without trade disputes. No other animal protein commodity is as actively traded internationally between countries. For example, in 1999 the United States imported more than 50% of its seafood consumption products from 172 countries (valued at US$9.0 billion) and exported seafood to 162 countries (NMFS 2000). In 1999 the value of US seafood exports was only US$2.8 billion, resulting in an edible fishery trade deficit of US$6.2 billion, of which approximately US$3.1 billion was attributed to shrimp (much of it aquacultured). In addition to the United States, other countries' import/export trade deficits, relative to their international trade in fishery products, can be computed from Tables 5.1 and 5.2. The top five countries exporting seafood to the United States were Canada, Thailand, China, Ecuador, and Chile. Preliminary estimates for 1999 indicate a reduction in world fish production following the substantial decline to 117 million mt in 1998 (FAO 2000). "El Niño" cannot be blamed for the totality of this reduction since shrimp exports from Ecuador fell by 22% between January and September 1999 because of

TABLE 5.1. Fishery Products Imports in Million US$

Country	1993–95 Average	1996	1997	1998
World Total	**50,704**	**57,217**	**56,727**	**54,823**
Japan	16,060	17,024	15,540	12,827
United States of America	6,825	7,080	8,139	8,579
Spain	2,791	3,135	3,085	3,546
France	2,858	3,194	3,062	3,505
Italy	2,223	2,591	2,572	2,809
Germany	2,227	2,543	2,363	2,624
United Kingdom	1,806	2,065	2,142	2,384
Denmark	1,361	1,619	1,521	1,704
China (Hong Kong SAR)	1,618	1,928	2,097	1,612
Netherlands	1,000	1,142	1,107	1,230
Canada	923	1,159	1,129	1,195
Belgium	896	966	979	1,061
China (excluding Taiwan Prov. & Hong Kong SAR)	791	1,184	1,183	991
Portugal	687	783	750	926
Norway	374	536	562	675
Sweden	455	587	596	639
Korea, Rep.	694	1,054	1,018	562
Thailand	824	818	866	814
China, Taiwan Province	565	613	660	482
Australia	406	475	483	442
Brazil	287	482	484	455
Singapore	615	642	627	416
Switzerland	387	396	361	387
Poland	161	240	262	304
Others	3,687	4,678	4,841	4,361

Source: FAO 2000.

white spot disease. Likewise, it is reported that only 30% of the shrimp farms in the Guaymas are operational because of shrimp disease. Additional losses of shrimp supplies can be expected as shrimp diseases advance throughout Latin America, Asia, and Oceania (FAO 2000).

Compounding the difficulty of international trade in seafood products is the fact that numerous countries and regional customs organizations, such as the EU, are implementing more sophisticated product quality and safety regulations, such as HACCP, to prevent, control, or minimize seafood hazards to an acceptable level of protection. Although it is estimated that 65% of the total international fish trade is performed under HACCP-based regulations, ensuring compliance with those regulations to meet harmonized, accepted, and transparent definitions, standards, and conformance assessment procedures will need further validation (FAO 2000). Aside from the human health considerations there are also aquatic animal and plant health considerations and environmental resource sustainability issues associated with

TABLE 5.2. Fishery Products Exports in Million US$

Country	1993–95 Average	1996	1997	1998
World Total	**46,875**	**52,824**	**53,317**	**50,972**
Norway	2,714	3,415	3,399	3,661
Denmark	2,323	2,699	2,649	2,898
China (excluding Taiwan Prov. & Hong Kong SAR)	2,233	2,857	2,937	2,656
United States of America	3,264	3,148	2,850	2,400
Thailand	4,015	4,118	4,329	3,971
Canada	2,184	2,291	2,271	2,265
Indonesia	1,557	1,678	1,621	1,628
Chile	1,378	1,697	1,782	1,597
China, Taiwan Province	1,921	1,762	1,780	1,580
Spain	1,008	1,461	1,471	1,529
Iceland	1,248	1,426	1,360	1,434
Netherlands	1,393	1,470	1,426	1,365
United Kingdom	1,137	1,308	1,264	1,549
Korea, Rep. of	1,437	1,509	1,376	1,246
Russian Federation	1,609	1,686	1,356	1,170
India	1,001	1,116	1,128	1,135
France	920	1,003	1,098	1,097
Germany	781	1,056	977	1,053
Others	10,471	13,159	13,428	12,227

Source: FAO 2000.

the international trade of aquacultured products that must be met. This chapter details the principal international organizations that specialize with the aforementioned issues and, in general, the principal manner in which each organization addresses these concerns.

WORLD TRADE ORGANIZATION

As indicated in Chapters 1, 2, and 3, the world trade of fish and fishery products is enormous and can represent a significant source of export income, particularly for developing countries. A substantial portion of the fishery export trade volume consists of aquacultured products.

The antecedent of the World Trade Organization (WTO) began in 1948 when the United Nations established the General Agreement on Tariffs and Trade (GATT). GATT served as the primary rule-making mechanism for much of the world trade during a subsequent time of phenomenal growth of international trade in commodities and services. Through the GATT process, it was envisioned that an institution known as the International Trade Organization (ITO) would be formed as a part of the United Nations.

Nearly one-half of the member governments agreeing to GATT consented to reduce or eliminate customs tariffs, stimulate trade liberalization, reduce nationwide protection measures, and accept, on a provisional basis, a portion of the draft charter requirements for the ITO that were being negotiated. Although the ITO charter was not ratified, which negated its mandatory standing, it maintained a provisional status until the WTO was subsequently established in 1995 (USTR 1994). Furthermore, GATT served as the only recognized multi-international instrument governing trade until 1995. GATT also provided the sponsorship and forums for a series of multilateral trade negotiations known as "trade rounds," from which significant achievements in international trade liberalization were made.

The early trade rounds focused on further reducing tariffs. The "Kennedy Round," held during 1964–1967, conducted tariff reduction activities and introduced antidumping measures. Antidumping measures are usually described as importing country measures taken against imports of a product when the export price of the product is below the price of the product in the domestic market of the exporting country and the dumped imports cause injury to a domestic industry in the importing country. Stated in another way, an antidumping measure is an importing country's economic action taken against an exporting country selling an exported product at an unfairly low price. During the Tokyo Round (from 1973 to 1979) of GATT negotiations, it was determined that "technical" barriers are an important category of nontariff trade "protectionist" barriers that can be faced by exporters. This recognition was the first international attempt to deal with "nontariff trade" barriers. Such nontariff trade barriers often take on the form of unjustified hygiene or safety requirements. Because the GATT Agreements and Rules were not subscribed to by the full membership during that time, these requirements were referred to as "Codes."

The Uruguay Round of GATT negotiations took place from 1986 to 1994. Not only were these negotiations the largest of any kind in history, they also resulted in the largest reform in world trading since GATT's inception. Since then WTO has subsumed, streamlined, and greatly expanded the earlier GATT requirements. The objective of the WTO is to assist international trade to flow smoothly, freely, fairly, and predictably. The WTO administers trade agreements, serves as a forum for trade negotiations, settles trade disputes, reviews national trade policies, assists developing countries in trade policy issues through technical assistance and training programs, and cooperates with other international organizations. As of November 2000 there were 140 members of WTO, representing more than 90% of the world's population. The WTO recognizes the least-developed countries (LDC) as acknowledged by the United Nations. These are a group of 48 developing countries that the United Nations has identified as LDC, owing to their poverty, weak human resources, and low level of economic diversification. Of these countries, 29 are members of the WTO.

Within the WTO process, developing countries must define themselves; however, country self-designation is not automatically accepted in all WTO matters. The designation of a developing country or LDC is of great significance in WTO because special privileges concerning longer time periods for implementing agreements and commitment to compliance measures and WTO decisions agreed to by developed countries can lead to increased trading opportunities. Furthermore, as a result of the Uruguay Trade Round, WTO members are required to safeguard the trade interests of developing countries by providing them with assistance to build the necessary infrastructures needed to participate in the WTO, handle disputes, and provide assistance to implement technical standards.

Decisions and agreements made within the WTO must be ratified by an individual national member country's legislative or parliamentary bodies. The ultimate decision mechanism within the WTO is the Ministerial Conference which meets at least once every two years. Figure 5.1 provides the table of organization for the WTO.

Reporting to the Ministerial Conference of the WTO is the General Council. This council serves as a trade policy review body, as well as a dispute settlement body. The General Council meets several times a year and consists of member country representatives to the WTO. Country Ambassadorial or Head of Delegation Rank signifies the political importance that member countries place on WTO endeavors.

Reporting to the General Council are three action councils: (1) Goods Council, (2) Services Council, and (3) Intellectual Property Council. Within the organizational structure of the WTO are numerous specialized committees, working groups, and working parties that deal with a myriad of issues such as individual agreements and environmental issues. Development of prospective members, regional trade agreements, trade and investment interrelationships, trade and competitive policies, and transparency in government procurement are other issues dealt with by the specialized committees and working groups.

As with most international organizations, the WTO is supported by a well-staffed Secretariat, based in Geneva, Switzerland, for technical support and assistance in the dispute settlement processes. The

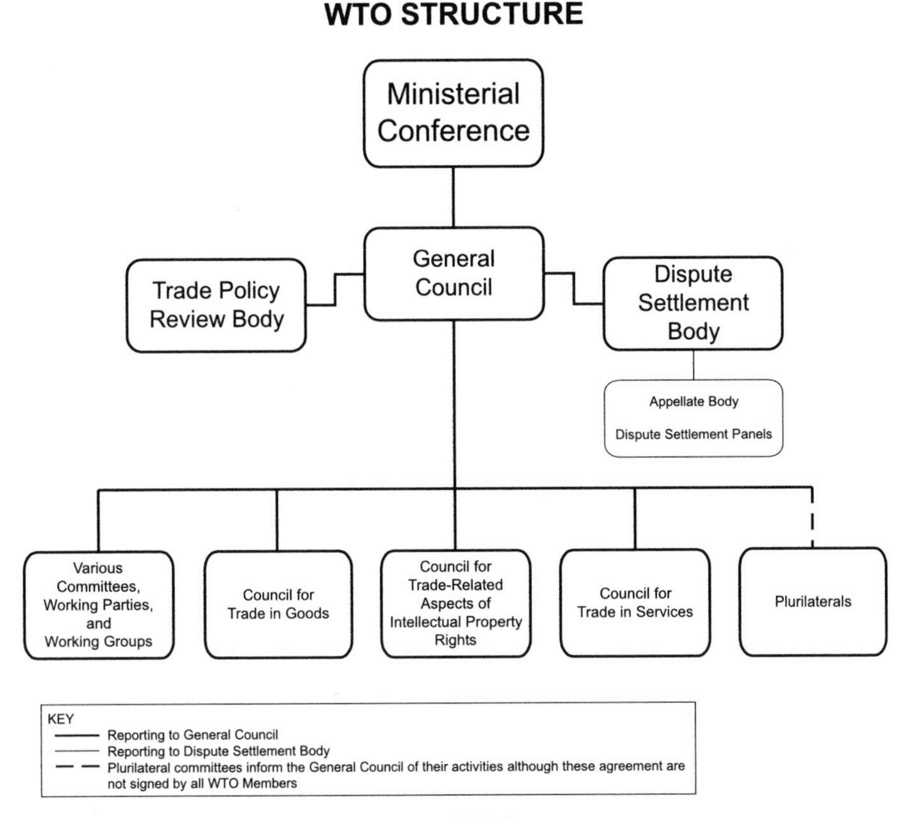

Figure 5.1. *WTO Structure*

Secretariat does not exercise a decision-making function, because all decisions of the organization must be determined and ratified by the member countries.

The three main overarching areas managed by the WTO relate to: (1) GATT, which addresses the trade of goods; (2) GATS, which relate to services supplied from one country to another, such as international telephone calls, consumers or firms making use of a service to another country, such as tourism, a foreign company setting up subsidiaries, such as aquaculture operations, banks, or individuals traveling from their own country to supply services in another, such as consultants, and (3) TRIPS, dealing with copyrights and rights related to copyrights such as literary and artistic works and industrial property considerations such as trademarks, geographical indicators, patents, industrial designs, and trade secrets. All of these agreements address a wide range of multilateral trading issues concerning agriculture (including fisheries), textiles and clothing, banking, telecommunications, government purchases, industrial standards, food sanitation regulations and intellectual property, as well as other trading issues. The WTO has some very distinct requirements in its General Principles governing the international trading of its member countries. These General Principles relate to trading without discrimination among or between a country's trading partners, such as not granting one trading partner a special concession, or other discriminatory actions. This requirement to trade without discrimination in WTO lexicon is referred to as the "most favored nation" (MFN) concept. This WTO principle includes the requirements of "national treatment," which obligates an importing country to treat imported products (once the import has entered the domestic market) and domestic products in the same manner.

Another WTO General Principle requires its members to promote fair trade through negotiations to lower trade barriers, such as reducing tariffs and import bans on quotas. A third General Principle affecting importing countries is the requirement to foster a predictable trade environment so that the international exporting community is assured of stable and predictable import/export conditions. This General Principle can relate to "bindings" or ceilings on import custom tariff rates, discouraging quotas, and necessitating that importing country requirements are "transparent," which means that they have been developed through a public process and are stated in a clear, concise, and easily understandable manner. Such a transparent regulatory development process ensures that the import/export business community readily understands importing country requirements and that they will not encounter "shrouded" nontariff trade barriers. This assurance provides for a stable and predictable environment among international trading partners, thereby accomplishing one of the fundamental objectives of the WTO.

A fourth WTO General Principle requires the promotion of fair trade by its members. It should be understood that WTO makes a distinction between "free" and "fair" trade. The various agreements that WTO administers allow for import tariffs or country duties, and in special circumstances, other forms of protectionistic activities. Therefore, the WTO is not, nor should it be considered, as a "free trade organization." Rather, it is a "fair trade organization" dedicated to developing and enforcing rules that allow for fair, undistorted, and nondiscriminatory importing and exporting of goods, services, and intellectual property between international trading partners.

The final WTO General Principle, which it inculcates in all of its agreements, is the expectation that members in all of their WTO activities will support sustainable development and economic reform. This general requirement recognizes that more than three quarters of the WTO members are either developing countries, or countries transitioning to market economies. Recognizing that developing countries are acceding to obligations required of developed countries, WTO agreements provide for a longer transitional time for developing countries to meet all WTO agreement requirements. LDC have even more flexibility in implementing the agreement's various requirements. The WTO, through its decision-making process, clearly expects developed countries to provide increased technical assistance and expedited market access to LDC.

It must be recognized that fishery resources, whether wild caught or of an aquacultured origin, are used for food purposes (acknowledging that some are used for aquaria or ornamental jewelry purposes). Therefore, aquacultured animals and products are subject to the usual and customary national and international food hygiene trade regulations. Garrett et al. (1998) indicated that there are two specific WTO agreements that are of particular specific significance to animals, plants, and food products including those of aquaculture origin. These are the aforementioned SPS and TBT Agreements. The SPS Agreement (referred to as a Measure) provides the basic rules for food safety and animal and plant health standards. The Agreement calls on countries to "further the use of harmonized measures . . . on the basis of international standards, guidelines and recommendations developed by relevant international organizations, which are the Codex International Food Standards Programme, the Office International des Epizootics (OIE), and the International Plant Protection Commission (IPPC)" (Garrett et al. 1998), often referred to as the "three sisters" of international standards-setting bodies. The SPS Agreement does not require countries to change "their appropriate level of protection" for life or health. It does, however, obligate a member to accept the SPS Measures of another member as equivalent to its own even if those measures differ, as long as they achieve the same level of sanitary and phytosanitary protection. The SPS text preserves a country's ability to maintain its own standards providing they are based on science and achieve the level of protection that the importing country deems appropriate.

Harmonization in this international trading context is defined as "the establishment, recognition and application of common sanitary and phytosanitary measures by different countries" (WTO 1995a). The SPS Agreement contains 14 binding articles and three annexes (Table 5.3) that address the requirements for regulation transparency and control,

TABLE 5.3. GATT Sanitary and Phytosanitary Agreement Articles and Annexes

• General Provisions	• Transparency	• Implementation
• Basic Rights and Objectives	• Control, Inspection, and Approval Procedures	• Final Provisions
• Harmonization	• Technical Assistance	• Annex A—Definitions
• Equivalence	• Special and Differential Treatment	• Annex B—Regulation Transparency
• Risk Assessment and Management	• Consultations and Dispute Settlements	• Annex C—Control, Inspection, and Approval Procedures
• Regional Considerations	• Administration	

inspection, and approval procedures. Substantial requirements are necessary to comply with these regulations.

The TBT has requirements corresponding to those of the SPS. The TBT features many activities dealing with standards development and conformity assessment considerations as well as a number of specific considerations dealing with international and regional system requirements. Provision of information, assistance, and technical support by developed country members to developing countries in preparing technical regulations is also a requirement of the TBT. The ability to provide differential and more favorable treatments to developing countries (WTO 1995b) and definition of a specific dispute resolution and settlement mechanism are other responsibilities of the TBT.

It is sometimes difficult to differentiate between a SPS and a TBT Measure. The SPS Agreement requirements are Measures directed toward protecting human or animal life from risks due to contaminants, toxins, food additives, or disease-causing organisms. SPS Agreements also contain criteria to protect animals or plant life from diseases, pests, or disease-causing organisms. The SPS Measures include means to prevent or limit other damage to a country resulting from the entry, establishment, or spread of pests, and they relate to the protection of aquatic animal fish health and wild fauna, as well as forests and wild flora.

The TBT Agreement covers all technical regulations, voluntary standards, and associated conformity assessment procedures used to determine compliance with these regulatory conditions, except when such requirements are specific sanitary or phytosanitary measures as defined by the SPS Agreement. Likewise, food labeling requirements, nutritional claims, product quality, and packaging requirements are also TBT issues in cases of dispute. However, regulations addressing food microbiological contamination, pesticide or veterinary drug residue levels, or permitted levels of food additives are, by definition, SPS considerations.

Governments may challenge another country's importing/exporting regulations. When considering such actions, it is necessary to know whether the measure in dispute is a SPS or a TBT violation, because the governmental burden of proof differs between the two agreements.

Both SPS and TBT challenges have been made for wild-caught fishery products and can be expected to be made for aquaculture products as well.

Selected WTO Fishery Trade Settlement Case Histories

When conducting global trade in fishery products, understanding how the national legislation and associated regulations relate to fishery management and/or environmental concerns pertaining to international trading requirements under GATT and WTO provisions is essential. For example, national governments (and associated private sector interests) often lose a WHO trade dispute because they do not understand the restrictive nuances of the international WTO requirements.

Before the establishment of WTO in 1995, a procedure for settling trade disputes existed under GATT. That procedure, however, had no fixed time lines, and rulings were easy to block because a consensus (including agreement by the country losing the dispute) was required. Therefore, dispute cases could drag on indeterminately, resulting in no conclusive ruling. The creation of WTO significantly improved the dispute settlement process in that now the time line for dispute settlement should take no longer than 15 months (including appeals), a formalized 14-step procedural process has been institutionalized, and a losing country can no longer block the adoption of the dispute panel settlement ruling.

For instance, in 1982 before the establishment of WTO, the USA brought forth a case against Canada to prohibit the importation of tuna and tuna products from Canada. This import ban against Canada was imposed by the United States after Canada had seized US fishing vessels and arrested their crews for harvesting albacore tuna without permission from the Canadian government in waters that were considered by the Canadian government to be under their jurisdiction but were not so recognized as such by the US government. The US import prohibition was introduced to WTO under its Fishery Conservation and Management Act as a retaliation measure against Canada for the US vessel seizure and crew arrest. The dispute panel ruled against the US position because the import prohibition was contrary to Article XI:1 and was not justified under Articles XI:2 or XX(g) of the General Agreement (GATT 1982).

Likewise, in 1988, the United States brought forth a case against Canada over the Canadian regulations prohibiting the exportation or sale for export of certain unprocessed herring and salmon. The Canadian export regulations were promulgated under the Canadian Fisheries Act, and the Canadian government argued that the export restrictions were an integral portion of the overall system of fishery resource management measures necessary to conserve fishery resource stocks. The dispute panel found that the measures mandated by Canada were contrary to the GATT Provisions contained in Article XI:1 and, furthermore, were not justified by Articles XI:2(b) or XX(g) (GATT 1988).

Between 1991 and 1994, a number of countries, led by Mexico and the EU, brought cases against the United States for its dolphin protection standards and associated regulations pursuant to the country's Marine Mammal Protection Act (MMPA). In the eastern tropical Pacific Ocean, schools of yellowfin tuna often swim beneath schools of dolphins. With purse seine harvesting of such tuna, dolphins can become trapped in the nets and die unless they are released. At issue was the fact that the United States required its domestic American fishing fleet, and countries whose fishing vessels harvested yellowfin tuna in the eastern tropical Pacific Ocean, to meet the incidental kill or serious injury provisions in relation to marine mammals contained in the USA Act. The USA import regulatory prohibition also applied to what are referred to as intermediary countries that might be handling the violative tuna, such as Costa Rica, Italy, Japan, France, Netherlands Antilles, Spain, and United Kingdom, before final export to the United States. Members of ASEAN were also named as intermediaries. These dispute cases became known as the "tuna-dolphin" and "son of dolphin-tuna" cases, and they continue to attract attention even though they were handled under the pre-WTO/GATT dispute settlement process, which required a consensus. These particular cases were significant because of the implication that through an import trade regulation, one country was implying to another country what national environmental regulations must be in place in the exporting country before importation of products will be allowed by the importing country. Furthermore, these cases raised the question, Do trade import rules permit action to be taken against the method used to produce goods, as opposed to quality of the goods? (i.e., a "process" vs. a "product" issue). The GATT Dispute Settlement Panel was restricted to determining only how GATT requirements applied to the issue and not whether the USA's policy was environmentally correct. The Panel was asked to determine whether the US requirement for tuna products to be labeled "dolphin-safe" was a GATT violation. Although the Panel ruled that, except for the "dolphin-safe" labeling provision, the other US import and embargo sanctions were violations of GATT provisions and not were justified by Articles XI:1, XX(b) and (g), and XX(d), they also indicated that through the GATT process, the US position could be made compatible with GATT provisions. The Panel's report, however, was not implemented (GATT 1991).

Since the formation of the WTO, two notable fisheries-related trade dispute issues have been brought before the WTO. The first dispute was a landmark case brought against the USA by India, Malaysia, Pakistan, and Thailand. This case became known as the "shrimp-turtle" case because it dealt with US technical regulations to prevent or minimize turtle by-catch in shrimp harvesting. The US Endangered Species Act of 1993 listed five species of sea turtles that occur in US waters as endangered or threatened and prohibited their take within the USA, its territorial seas, and on the high seas. This designation required US shrimp trawlers to use turtle excluder devices (TED) in their nets when fishing in waters where there was a significant likelihood of harvesting net interaction with sea turtles. Similarly, an import law was enacted by the United States that prohibited the import of shrimp or shrimp products from countries that harvested shrimp with harvesting net technology that might adversely affect the sea turtles of concern. Moreover, the United States required that the affected shrimp-exporting countries desiring to import shrimp products must place into effect similar TED technology regulations and requirements that the United States was imposing on its own domestic fleet. The United States lost the shrimp-turtle case because it discriminated between WTO members (trading partners) by providing technical and financial assistance, and longer transitional periods to use TED to Caribbean countries and their fisheries, than it did to the Asian countries of India, Malaysia, Pakistan, and Thailand, which filed their own complaint with the WTO. The WTO Trade Dispute Panel ruling clearly indicated that the United States was well within its sovereign national rights to exercise protection and preservation of the environment, as well as adopting effective measures to protect endangered species such as sea turtles. The Panel also concluded that when enacting such measures, the USA discriminated between Carribean and Asian developing countries in that import regulations cannot be applied in a manner that provides for arbitrary or unjustifiable discrimination between countries (trading partners) where the same conditions prevail, or as a disguised restriction on international trade. The WTO ruling further indicated that members are free to adopt their own policies for protecting their environment as long as in so doing they fulfill their obligations and respect the rights of other members under the WTO agreement (WTO 1998a).

Another WTO trade dispute settlement case of interest (because it can more directly relate to aquaculture pursuits) is the case brought by Canada against Australia for the prohibition of the importation of raw or unheated salmon. In this case the EC, India, Norway, and the United States governments also made third-party submissions to the Dispute Panel relative to the issue. Canada claimed that the Australian action was an illegal import prohibition under Article XI:1 of GATT 1994 and, furthermore, was not justified under Article XI:2 or Article XX in GATT 1994. Canada made seven additional claims against Australia relating to assertions:

1. that Australia did not follow Article 3.1 of the SPS Agreement because its action was not based on existing OIE standards;
2. an improper salmon health risk assessment was conducted;
3. pertinent available scientific evidence was not considered as required by the SPS Agreement;
4. the action represented an arbitrary and unjustified distinction in levels of protection and also represented a disguised restriction of trade;
5. the action or measure was more restrictive than required to meet Australia's necessary level of protection; the measure violated Article 2.2 of the SPS Agreement because it was maintained without sufficient scientific evidence;
6. the measure arbitrarily or nonjustifiably discriminated between members when similar conditions prevailed; and
7. Canada had suffered huge economic losses from the Australian action.

Australia denied the Canadian claims, making counterarguments to each and every charge. A WTO Dispute Panel report was published in 1998, which found in behalf of the Government of Canada (WTO 1998b).

CODEX ALIMENTARIUS COMMISSION

During the late 1800s and the early 1900s, the Austro-Hungarian Empire promulgated a number of voluntary food and product standards entitled "Codex Alimentarius Austriacus," whose Codes were used by reference courts to determine standards of identity for foods (FAO/WHO 1999a). The current Codex Alimentarius International Food Standards Programme (Codex) draws its name from that early Austrian approach. WTO recognizes that Codex is an important international food standards reference point for foods in international trade. Adherence to Codex voluntary standards can provide a "legal safe harbor" in trade disputes brought before the WTO, thus giving Codex increased prominence (Garrett et al. 1997).

Codex was founded by the United Nations in 1962 after the FAO discovered that there were 135 different organizations working on various aspects of international food standards and related issues. The Codex Programme is sponsored by both the WHO and the FAO. Currently there are more than 160 countries participating in this massive international effort. Actions of the Codex Programme greatly influence global regulatory food control activities because Codex work products represent a consensus of opinion on regulatory issues by member countries, which in turn represent more than 97% of the world's population. The Codex Programme is operated by an International Commission through an Executive Committee and has various subsidiary bodies. Figure 5.2 depicts the organizational structure of the Codex Alimentarius Commission (CAC) and its various Subsidiary Bodies. These Subsidiary Bodies or Committees are both vertical and horizontal (or crosscutting in nature). For example, specific food commodity committees, such as the Codex Committee on Fish and Fishery Products, represent vertical committees. The Codex Committee on Food Hygiene, which addresses hygienic considerations in all of the outputs of the Codex Alimentarius Programme, is an example of a horizontal or crosscutting committee. Additionally, there are Regional Committees, are also crosscutting in nature, which address special needs of specific geographical regions. Along with member nations, Codex relies on scientific support from three prestigious committees sponsored by other specific United Nations Programmes. These committees are the Joint Expert Committee on Food Additives, the Joint Meeting on Pesticide Residues, and the International Consultative Group on Food Irradiation (which may soon be *sine die*, because its principal work has been accomplished). In addition, Codex also relies on three separate expert intergovernmental task forces to advise on specific issues

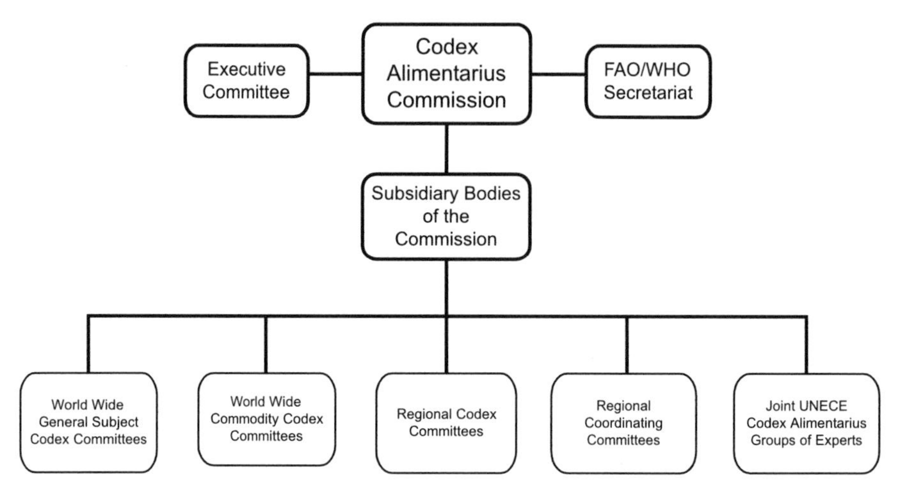

Figure 5.2. *Codex Organization*

dealing with foods derived from biotechnology, animal feeding, and fruit and vegetable juices.

The products of Codex generally relate to the following four specific areas: (1) the development of General Principles to be followed in the international trade of food commodities, (2) specific Codex Commodity Standards for individual food commodities, or processing requirements, (3) the establishment of Codex Guidelines for specific actions or procedures, and (4) recommended Codes of Hygienic Practice, which are similar to the USA's GMP concept that is to be followed when producing and/or manufacturing specific food commodities (Garrett et al. 1998).

Codex Commodity and General Standards emphasize the need for priority consideration to be given to food safety, hygiene, and other concerns to ensure that the foods consumers receive through international trade, when produced and inspected by Codex Standards and Codes, will be safe, do not present health hazards, and are of a minimal acceptable quality level. The rationale and types of processes for setting such standards have been described in detail by Dillon and Griffith (1997). The format for each commodity standard contains 10 provisions to ensure that the marketing of the food is not misleading and that food items are properly labeled relative to their contents (FAO/WHO 1997a). The Codex Standard document format contains the name of the standard, its scope, description of the product or product's definition, including raw material condition (if appropriate) and types and styles of product and pack; weights and measures; labeling; and methods of analysis and sampling. The Commodity Standard format also addresses essential composition and quality factors to ensure a minimally acceptable product and contains provisions for dealing with food additives, contaminants and hygiene requirements, each of which are singularly focused at delineating food safety requirements.

Apart from Commodity Standards, Codex elaborates General Standards that have cross-cutting application to all foods, and are not product specific. There are General Standards for:

- Ethics for international trade in foods
- Food labeling
- Food additives
- Contaminants
- Methods of analysis and sampling
- Food hygiene
- Nutrition and foods for special dietary uses
- Addition of essential nutrients to foods
- Food import and export inspection and certification systems
- Residues of veterinary drugs in foods
- Pesticide residues in foods

One can observe from the titles of these General Standards that they are specifically focused to meet the Codex mission to protect consumers and facilitate trade. As indicated above, Codex has been a science-based activity since its founding. The Programme provides a forum for the world's leading experts to discuss, debate, and reach a scientific consensus on the food safety issues that affect international trade. Furthermore, governmental participation allows access to the world's most current and complete body of scientific food safety information. The Programme has provided a focal point for food-related scientific research and investigation and has stimulated research and developments in food technology, microbiology, and pesticide and veterinary drug residues. Much of the Codex work is accomplished through collaborative studies between individual scientists or institutions such as government laboratories, research, and/or academic institutions or by joint WHO/FAO Expert Committees and Consultations. Selection criteria for individuals to participate in WHO/FAO Joint Expert Consultations include recognition that the participating individuals (1) are eminent independent scientists in their specialty, (2) have the highest respect of other scientific peers, (3) are impartial and indisputably objective in their judgement, and (4) can accept appointment to the Consultation as individual scientists and not as government representatives or spokespersons for organizations. Furthermore, geographic balance of expertise is also taken into consideration for participation (Burke 2000). Recent expert consultations addressing food safety issues associated with products from aquaculture relate to control of foodborne trematode infections (WHO 1995a) and food safety issues associated with products from aquaculture (WHO 1999). Other expert consultations have general application to all food products including those from aquaculture. These include Application of Risk Analysis to Food Standards Issues (FAO/WHO 1995), Guidance on Regulatory Assessment of HACCP (FAO/WHO 1998), Guidelines for Strengthening a National Food Safety Programme (WHO 1996), Joint Group of Experts on the Scientific Aspects of Marine Environmental Protection (FAO 1997), and Training Aspects of the Hazard Analysis Critical Control Point Systems (HACCP) (WHO 1995b).

Furthermore, FAO, through its regular fishery resource programs, develops similar Codes of Practice for wild harvest and aquaculture commodities such as the Code of Conduct for Responsible Fisheries—Aquaculture Development (FAO 1995), which includes specific aquaculture considerations.

It cannot be overemphasized that sound science plays a preeminent role in the Codex decision-making process to ensure that the organization's standards and recommendations can withstand the most rigorous scientific scrutiny. It is for this reason that Codex has a formal mandate governing the role of science throughout its decision-making process as follows:

STATEMENTS OF PRINCIPLE CONCERNING ROLE OF SCIENCE IN THE CODEX DECISION-MAKING PROCESS AND THE EXTENT TO WHICH OTHER FACTORS ARE TAKEN INTO ACCOUNT

1. The food standards, guidelines and other recommendations of Codex Alimentarius shall be based on the principle of sound scientific analysis and evidence, involving a thorough review of all relevant information, in order that the standards assure the quality and safety of the food supply.
2. When elaborating and deciding upon food standards Codex Alimentarius will have regard, where appropriate, to other legitimate factors

relevant for the health protection of consumers and for the promotion of fair practices in food trade.

3. In this regard it is noted that food labeling plays an important role in furthering both of these objectives.
4. When the situation arises that members of Codex agree on the necessary level of protection of public health but hold differing views about other considerations, members may abstain from acceptance of the relevant standard without necessarily preventing the decision by Codex (FAO/WHO 1997b).

The completion of a Codex Proposed Draft Standard or other organizational outputs requires eight steps that may take several years to complete, to allow all countries and committees to fully review the proposed standard. Documents must first gain acceptance in two ways before they are adopted as Codex Standards. First, independent scientific experts from around the world review the scientific validity and soundness of proposed Codex Standards, Codes, Guidelines, or acceptable limits. Second, committees of regulators (who are very often subject matter experts themselves) must reach a consensus on the practicality and enforceability of the standard's application, limits, etc.

Unequivocally the work of Codex has been profound. Since its inception, the program, through a science-based consensus approach, has promulgated 237 Commodity Standards, including those for various types of processed fruits and vegetables; meat and fish products; cereals, pulses, and legumes; fats and oils; milk and milk products; soups and broths; and foods for special dietary uses. In addition to Codex Standards, 41 Codes of Hygienic Practice on technological practice have been published and 185 pesticides, 1005 food additives, and 54 veterinary drugs have also been evaluated. Limits for 3,274 pesticides have been set, and there 25 Guidelines for Contaminants have been established (Burke 2000). These achievements are truly remarkable.

Without a doubt, Codex has upgraded global food manufacturing practices that have resulted in dramatically improved global consumer protection. Fishery products are traded in the global marketplace between and among hundreds of countries. With such an expanding global economy it is rapidly becoming apparent that ensuring food safety is a massive and incredibly complex task, with no one country being able to accomplish the job alone (Garrett et al. 1994). It is necessary that all who produce, process, source, handle, serve, and consume food products around the world understand that because of the expanding world trade of food products, coupled with new sophisticated production and processing techniques, it is no longer sufficient to only establish and enforce domestic standards, but appropriate standards and practices must be developed and followed worldwide. To do less in the changing global paradox of increasing international food trade can lead to a false sense of security in the soundness of a country's food supply (Garrett 2001a). Codex provides an excellent vehicle to develop and promulgate those necessary standards. Many developed and developing countries have adopted Codex Standards in their national food code. Codex, through its activities, has lifted the world's food safety community awareness of food safety and related issues to unprecedented heights and has consequently become the most important international reference point for developments associated with food standards. It is a transparent government-to-government process that allows for international Nongovernmental Organizations (NGO) participation. In meeting its mission to either develop or coordinate regional food standards, Codex has had remarkable successes. Examples of such regional achievements include references to Codex activities in the North American Free Trade Agreement (NAFTA), Southern Consumer Market (MERCOSUR), and the Asia-Pacific Economic Council (APEC). NAFTA relates to Canada, the United States, and Mexico. MERCOSUR involves Argentina, Brazil, Paraguay, and Uruguay. APEC relates to promoting open trade in economic cooperation among 21 member countries around the Pacific Rim.

NAFTA includes two ancillary agreements dealing with SPS measures and TBT. With regard to SPS measures, Codex standards are cited as basic requirements to be met by the three NAFTA member countries in terms of the health and safety aspects of food products. MERCOSUR's Food Commission has also recommended a range of Codex standards for adoption by its member countries and is using other Codex standards as points of reference in continuing deliberations. APEC has drafted a Mutual Recognition Arrangement on Conformity Assessment of Foods and Food Products. This Agreement calls for consistency with SPS and TBT requirements as well as with Codex Standards. In addition, EU directives frequently refer to the CAC as a substantial basis for their requirements.

Recent Codex Products of Interest to Aquaculture

Recent Codex Standards and Guidelines are (1) General Principles for Food Hygiene, including the General Principles and Guidelines for HACCP, and (2) General Principles and Guidelines for Application of Microbiological Risk Assessment and Microbiological Criteria, which defines "transparent" as "Characteristic of a process where the rationale, the logic of development, constraints, assumptions, value judgements, decisions, limitations and uncertainties of the expressed determination are fully and systematically stated, documented, and accessible for review" (FAO/WHO 1999b).

In the author's opinion, this definition for transparent is a major advance in not only how risk assessment and management activities should be undertaken but also how the documentation elements of those activities should be conducted, so that there can be a clear understanding of the scientific basis on which the risk assessment and subsequent regulatory risk management requirements are premised (Garrett 2001b).

Other Codex products include Guidelines for Generic Official Certificate Formats and the Production and Issuance of Certificates; Guidelines on the Judgement of Equivalence of Sanitary Measures Associated with Food Inspection and Certification Systems; and Guidelines for the Exchange of Information in Food Control Emergency Situations.

OFFICE INTERNATIONAL DES EPIZOOTICS

The Office International des Epizootics (OIE) is the WHO's program for animal health and is the second of three international health organizations that promulgate standards, which when conformed to, can provide a legal safe harbor in cases of WTO trade disputes. The OIE was established in 1924, and by March of 2001 it consisted of 157 member countries.

The mission of the OIE is to inform governments of the occurrence and course of animal diseases globally and of the methods that can be implemented to control such diseases. The organization also coordinates international studies for surveillance and control of animal diseases and harmonizes regulations for trade in animals and animal products among member countries.

The governing body of the OIE is an international committee consisting of permanent delegates designated by the member countries. The activities of the organization are conducted by a Secretarial structure entitled "Central Bureau," whose chief executive, a "Director General," is appointed by the International Committee. The organization's headquarters are located in Paris, France. Figure 5.3 is a table of organization of the OIE. Of particular note is that the OIE, through its specialist commissions and subsidiary working groups, is active in fish and wildlife diseases.

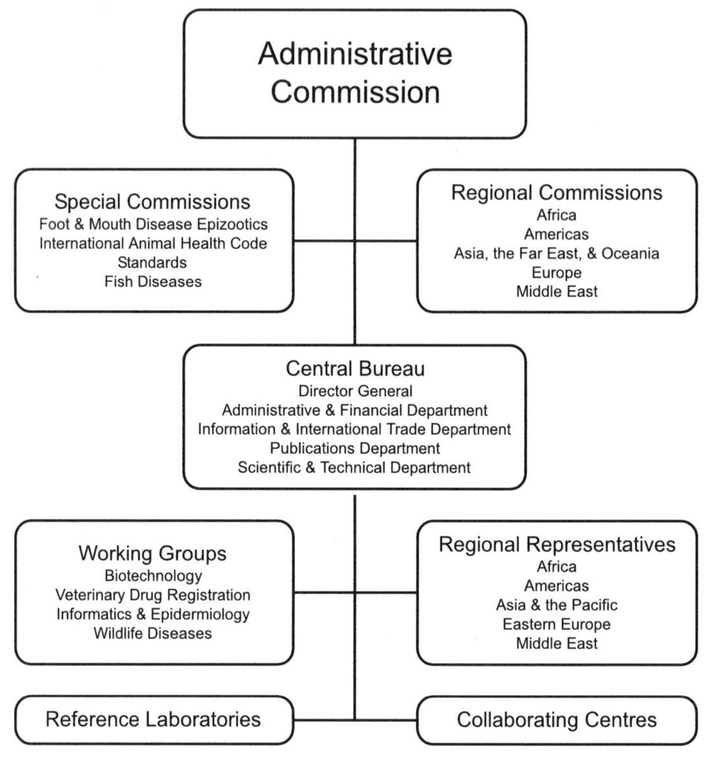

Figure 5.3. *OIE Structure*

The OIE requires that it be notified by member countries when certain diseases, or the agents which cause them, occur within their countries. Such diseases must be of a transmittable nature and are differentiated by their severity and spread. Aquatic animal diseases are listed in the International Aquatic Animal Code (OIE 2001) in two categories as "diseases notifiable to the OIE" and as "other significant diseases'. While member countries are not required to notify OIE upon finding an "other significant" aquatic animal diseases, they often share this information with OIE so that it is communicated to the other members. Both the notifiable diseases and other significant diseases are defined and listed in Table 5.4.

During the 2000 68th Annual General Session of the OIE, the increasing interest in aquatic animal health issues was expressed in the organization's deliberations. A summary of the organization's efforts on the prevention and control of aquatic animal diseases was presented at the session as follows:

> It was recognized that in order to prevent or reduce the risk of introduction of serious aquatic animal diseases and thus avoid economic losses in the aquaculture industry and in wild stocks, it is important to have a set of principles for the prevention and control of such diseases. These principles include the establishment of a legislative framework, both nationally and internationally. The key international documents to be considered are the OIE International Aquatic Animal Health Code and Diagnostic Manual for Aquatic Animal Diseases, as well as other documents, such EC Directive 91/67/EEC that applies health regulations to all European Union countries. Guidelines for management of diseases in Asia are currently being established by FAO/NACA, partly based on the OIE standards.
>
> In addition to a legislative framework, the most important factors involved in improved surveillance and prophylaxis of aquatic animal diseases include: listing of diseases in order of importance, procedures for inspection and control, import regulations, quarantine measures, procedures for the introduction of new species, transport regulations and restriction on movements, disinfection procedures, contingency plans, personnel training, and disease prevention at the aquaculture establishment level with regard to water treatment, vaccination, medical treatment, as well as hygienic and sanitary measures (OIE 2000).

A perusal of the OIE's organizational structure, deliberative processes, products, and past organizational successes indicates that it is well equipped to play a major role in the prevention, minimization, or control of spread of aquaculture animal diseases resulting from the international trade of fishery products. OIE's role can be expected to expand in the near future as the increasing impact of aquatic animal diseases becomes more globally recognized relative to possible unintended consequences resulting from the importing and exporting of diseased aquatic animals or products derived thereof.

Aquacultured animal health issues relating to international trade are becoming more prominent. For example, it has been reported (FAO 2000) that the white spot virus continues to play havoc with Central American shrimp farming. Ecuador experienced a US$500 million loss in export earnings in 1999 from its 1998 revenues because of white spot virus. This not only resulted in a huge shrimp product economic loss but also increased unemployment by 40% of the affected aquaculture workers. Likewise in 1999, because of white spot virus outbreaks, only 30% of the shrimp farms in the Guaymas were operational. Such disease outbreaks have been reported in northern Peru, where a considerable number of shrimp ponds are located.

TABLE 5.4. OIE Classification of Aquatic Animal Health Diseases

Diseases Notifiable to the OIE			
Definition	Crustaceans	Fish	Mollusks
Transmissible diseases that are considered to be of socio-economic and/or public health importance within countries and that are significant in the international trade in aquatic animals and aquatic animal products. Reports of these diseases are normally submitted once a year, although more frequent reporting may be necessary in some cases to comply with Articles 1.2.1.2 and 1.2.1.3 of the 2001 International Aquatic Animal Health Code.	Taura syndrome White spot disease Yellowhead disease	Epizootic hematopoietic necrosis Infectious hematopoietic necrosis *Oncorhynchus masou* virus disease Spring viremia of carp Viral hemorrhagic septicemia	Bonamiosis (*Bonamia ostreae, B.* sp.) Haplosporidiosis (*Haplosporidum costale, H. nelsoni*) Marteiliosis (*Marteilia refringens, M. sydneyi*) Mikrocytosis (*Mikrocytos mackini, M. roughleyi*) Perkinsosis (*Perkinsus marinus, P. olseni*)
Other Significant Diseases			
Definition	Crustaceans	Fish	Mollusks
Diseases that are of current or potential international significance in aquaculture, but that have not been included in the list of diseases notifiable to the OIE	Baculoviral midgut gland necrosis	Channel catfish virus disease	None at present

because they are less important than the notifiable diseases, or because their geographic distribution is limited, or it is too wide for notification to be meaningful, or it is not yet sufficiently defined, or because the aetiology of the diseases is not well enough understood, or approved diagnostic methods are not available.	Nuclear polyhedrosis baculovirouses (*Baculovirus penaei, Penaeus monodo-*type baculovirus)	Viral encephalopathy and retinopathy
		Infectious pancreatic necrosis
	Infectious hypodermal and hematopoietic necrosis	Infectious salmon anemia
	Crayfish plague (*Aphanomyces astaci*)	Epizootic ulcerative syndrome
	Spawner-isolated mortality virus disease	Bacterial kidney disease (*Renibacterium salmoninarum*)
		Enteric septicemia of catfish (*Edwardsiella ictaluri*)
		Piscirickettsiosis (*Piscirickettsia salmonis*)
		Gyrodactylosis (*Gyrodactylus salaris*)
		Red sea bream iridoviral disease
		White Sturgeon iridovial disease

Source: OIE 2001

INTERNATIONAL PLANT PROTECTION CONVENTION

The International Plant Protection Convention (IPPC), which is currently administered through an Interim Commission on Phytosanitary Measures, is the third "sister" of international organizations providing a legal framework of Standards and Guidelines that, when adhered to, can also provide a legal safe harbor before WTO Trade Dispute Settlement Panels. The IPPC is a multi-international treaty that is administered through FAO. The purpose of the convention is ". . . international cooperation in controlling pests and plant products and in preventing their international spread, and experiencing their introduction into endangered areas. . . ." (IPPC Preamble). As of 2000 the IPPC had 110 nations participating as contracting parties.

The IPPC was adopted by FAO in 1951, took force in 1952, and was amended in 1973 and 1997. The 1997 revisions positioned the organization to assume the increased international role it was given by the GATT Uruguay Round/SPS Agreement.

The 1997 revision to the IPPC instituted an Interim Commission on Phytosanitary Measures (ICPM) to provide for a clear focus and oversight body for the establishment of International Standards for Phytosanitary Measures (ISPM). The Interim Commission (which will eventually become institutionalized) engages in extensive examinations of the worldwide status of plant protection activities and also provides direction to the IPPC Secretariat's work program including ISPMs. The IPPC Secretariat duties include the coordination of the IPPC work program, especially those activities that result in the elaboration of the ISPMs. Within the IPPC, there is substantial involvement by national plant protection organizations (NPPO) and regional plant protection organizations (RPPO). From a regulatory perspective, the IPPC Intermediate Commission is not as institutionally mature as Codex or the OIE in terms of standards development or mandatory pest or disease reporting. Nevertheless, because of its new WTO responsibility, the IPPC and its newly formed Interim Commission can be expected to quickly progress in those areas.

Although the IPPC is a legally binding agreement, as with Codex, the Standards developed are only of a voluntary nature. Nevertheless, the voluntary nature of the Standards is not diminished in significance during the WTO trade dispute settlement process. The IPPC also has an internal dispute settlement process by which contracting parties may raise challenges to measures. This internal settlement dispute process is "nonbinding," but its results can be expected to have significant impact when its findings are incorporated into the binding WTO trade dispute settlement process. Since 1995, there have been 10 ISPMs dealing with:

ISPM 1	Principles of Plant Quarantine as Related to International Trade
ISPM 2	Guidelines for Pest Risk Analysis
ISPM 3	Code of Conduct for the Import and Release of Exotic Biological Control Agents
ISPM 4	Requirements for the Establishment of Pest-Free Areas
ISPM 5	Glossary of Phytosanitary Terms
ISPM 6	Guidelines for Surveillance
ISPM 7	Export Certification System
ISPM 8	Determination of Pest Status in an Area
ISPM 9	Guidelines for Pest Eradication Programmes
ISPM 10	Requirements for the Establishment of Pest-Free Places of Production and Pest-Free Production Sites

Future activities of the IPPC can be expected to focus on addressing and developing Standards for Biosafety, Genetically Modified Organisms, Invasive Species, Wood Packaging Material, Official Control Procedures, Pest Reporting and Systems Approaches to Risk Management, Laboratory Accreditation, Testing and Certification Procedures, and Approval Procedures for Post Entry Quarantine Facilities.

The primary plants produced through aquaculture efforts are seaweed products. These products are mainly red, brown, or green algaes that are used either for direct human consumption or for their hydrocolloidal properties (e.g., agar, carrageenan, and alginates) (Rudolph 2000).

Gracilaria and *Gelidium* represent the first historical commercial use of red algae (excluding edible algae). The genus *Gracilaria*, which contains more than 200 species, is harvested principally in Namibia, Japan, Thailand, Taiwan, and Vietnam and is used for both human production and agar production. Cocultivation of *Gracilaria* species in shrimp and fish farming ponds in Asia has been a common practice. Likewise, *Gracilaria* has been successfully cultivated in Chile after the collapse of the wild populations because of overexploitation. In fact, some aquaculture techniques employed in Chile for *Gracilaria chilensis* have been successful for augmenting wild populations that have been overexploited. Other culturable marine plants include specific species of *Porphyra*, *Kappaphycus*, *Eucheuma*, and *Chondrus*. For a more detailed description of red and brown algae seaweed products of economic significance, the reader is directed to the overviews provided by Rudolph (2000) and Llaneras (2000).

Recognizing that aquaculture plants can be diseased, the WTO will rely on the IPPC as the international organization to provide standards, guidelines, and expert advice in cases of trade disputes resulting from allegations of failure to follow the SPS Agreement.

It may well be expected that to the extent that plant diseases, pests, or invasive species may negatively impact seaweed production the IPPC will eventually promulgate those standards.

INTERNATIONAL STANDARDS ORGANIZATION

ISO is not an acronym but rather the organization's short title derived from the Greek word "isos," which means equal. ISO is the largest of the international standards-setting bodies. It is an NGO that was established in 1947 and is headquartered in Geneva, Switzerland. ISO is an international federation of national standards-setting bodies from 140 countries. The three categories of membership within ISO are: (1) Member body (a national body that must be representative of the principal standards-setting activity in a country); (2) Correspondent member (usually an organization in a country that does not have a fully developed national standards development activity); and (3) Subscriber Member (for countries with very small economic status). While Correspondent and Subscriber Members do not have an active role in technical and policy development they are nevertheless kept informed about matters of interest to them.

The magnitude of ISO is staggering to the uninitiated because the purpose of the organization is to "standardize" all goods and services to make procedures simpler thus increasing reliability and effectiveness. The scope of ISO covers standardization in all fields except electrical and electronic engineering standards. The technical activities of the organization are carried out in some 2850 technical committees, subcommittees, and working groups. Each of these groups consists of industry, research institutions, governmental authorities, consumer NGO, and other international organizations to resolve global standardization problems. As many as 30,000 experts participate in ISO-sponsored meetings annually (ISO 2001).

Several outputs from the ISO process are of particular significance to private aquaculture pursuits and associated governmental regulatory bodies. These outputs relate to the ISO 9000 and 14000 Series of Standards; ISO/IEC Guides pertinent to certification, registration, and certification; and the Standards developed by ISO Technical Committee 69 (statistical sampling). The ISO 9000 Series provides generic quality management elements, standards, and auditable criteria for assessing the adequacy of quality assurance programs (Dillon and Griffith 1997). The 9000 series basically requires a company to document what it does—and to do what it documents. There are five Standards in the ISO 9000 Series (ISO 9002, ISO 9003, and ISO 9004) (Peach 1992). ISO 9000 and ISO 9004 are Guidance Standards, in that they contain descriptive rather than proscriptive requirements. ISO 9001, ISO 9002, and ISO 9003 contain proscriptive requirements and are referred to as Conformance Standards. In the author's experience, producing products and/or services in accordance with the ISO 9000 Series requirements is challenging but when properly achieved, results in a much better product, reduces the economic drains associated with "price of nonconformance," (Crosby 1984) and increases customer satisfaction. ISO product standards relating to aquaculture include piping materials, pumps, valves, and many others. As aquaculture enterprises become more sophisticated and automated, increased awareness and utilization of the ISO 9000 Quality System Management Standards to improve process management can be expected to occur. The project management processes and benefits for industry adhering to the ISO 9000 Standards were well described by (Badiru 1995).

ISO 14000

During the Marrakesh meeting of 1994, the GATT Ministers signed the final act that institutionalized the results of the Uruguay Round of GATT. Among other things, this directed the first meeting of the General Council of the WTO to establish the Committee on Trade and Environment, which would be open to all WTO members and would address the relationship between trade measures and the environmental measures necessary to promote sustainable development and for other purposes. The formation of this specific committee gave recognition to the idea that with the growing world economy and international trade, such activities can be expected to be accompanied by environmental degradation unless sufficient effective environmental management systems, safeguards, and conformance assessment procedures are set in place to ensure sustainable development.

Since the advent of WTO, and its TBT, conformity assessment to ISO Standards has assumed greater importance. As a result, ISO has published an essential reference entitled "Conformity Assessment: Guides and Standards." This ISO publication is a collection of trade-facilitating documents that has been developed within the ISO Committee on Conformity Assessment. ISO14000 Standards are included and were designed to meet the needs of business concerns as well as those of other parties interested in conducting conformance assessment audits of environmental management systems.

The ISO14000 Standards are designed to support WTO objectives of sustainable development and can be implemented in any type of organization within either the public or private sector. The ISO 14000 family of standards falls within the purview of the ISO Technical Committee 207—Environmental Management Committee. An ISO 14000 Case Study review of Models for Implementation published in 1996 describes the experiences of 25 companies (ISO 1996).

ISO Technical Committee 69—Application of Statistical Methods

ISO Technical Committee 69 produces standards that have a profound significance for organizations and/or products engaged in international trade. This ISO Technical Committee has four subcommittees that deal with terminology and symbols, statistical process management and control, acceptance sampling, and metrology. The importance of the ISO statistical methods Committee and Subcommittees' activities is that they develop statistical reference methods.

Often in national and international trade, buyers, sellers, or import/export regulatory control authorities can each sample a specific lot of product (each utilizing the same sampling plan) and get different results, thereby each concluding a different opinion on the product's status relating to food safety or economic value (product quality). In such cases the different opinions frequently result in legal suits or international trade disputes. Often the reason for the disputes arises from a lack of understanding by each of the parties of the nuances of statistical methods incorporated into regulatory or third-party conformance assessment procedures utilized by the import/export business community or regulatory control authorities. ISO Technical Committee 69 addresses applicable statistical methods for such matters and further provides numerous sampling plans/systems/schemes having various sensitivities in detecting possible nonconformances. These procedures provide for evaluation of the probabilities of accepting nonconforming product and not accepting conforming product. Failure to understand the statistical ramifications of sampling plans can lead one to a false sense of security when they are employed (Rainosek 1997).

ORGANIZATION FOR ECONOMIC COOPERATION AND DEVELOPMENT

The Organization for Economic Cooperation and Development (OECD) is a collection of 30 like-minded countries that provides a forum for governments to discuss, develop, and perfect economic and social policy. Its origin dates back to the European Economic Cooperation Organization (EEOC), which was formed to administer North American foreign aid, which under the Marshall Plan was used after World War II for the reconstruction of Europe. The OECD is at the forefront of determining how regional or individual countries' societal and economic policies can be expected to impact globalization of trade in goods and services and how expansion of world trade will change world economic growth or result in increased national protectionism.

Although the OECD addresses a wide range of issues, currently two refer to aquaculture. The first is the OECD's Committee on Fisheries, whose work addresses the interrelationships of competing users of the coastal zone in terms of development, wild-harvest fisheries, and aquaculture. This work was published in 1996 and gives a review of several country experiences relative to resource user conflicts in coastal zone areas and new coastal zone considerations that occur because of changing fisheries scenarios (OECD 1996). The OECD Committee on Fisheries also has been active in determining trade policy impacts resulting from direct or indirect fishery subsidies. This committee continues its efforts in reviewing fisheries (including aquaculture) in OECD countries as well as determining the economic impact when countries transition to responsible fisheries by using the aforementioned FAO Model for Responsible Fisheries.

Other investigations and reviews by the OECD, which will soon impact aquaculture pursuits, are the organization's efforts in addressing the many facets of biotechnology and genetically modified organisms in terms of food safety. The organization is conducting a review of national food safety systems, the harmonization of regulatory oversight in biotechnology and its impacts and possible benefits in biotechnology for developing countries. Although internationally it is well recognized that the Codex is the appropriate international body to finalize decisions in food safety (including biotechnology), OECD recognizes that it and other international forums can enrich the Codex debate. Therefore, as increased genetic manipulation occurs in aquaculture fishery and plant species, the OECD can be expected to provide an increasingly important international forum to debate and consider the associated societal ramifications including those related to public, aquatic animal, and environmental health concerns.

FOOD AND AGRICULTURAL ORGANIZATION OF THE UNITED NATIONS

The FAO was founded in 1945 as a principal United Nations organization. Its purpose is to raise the levels of nutrition and standards of living throughout the world by improving agricultural productivity (including wild-caught and aquaculture fishery pursuits) and bettering the quality of life conditions of rural populations. FAO currently has 180 member countries and one member organization, the EC. Since FAO's inception, its activities and programs have been responsible for quantum increases in food production resulting in dramatically reducing the proportion of hungry populations. One of FAO's specific priorities is to promote sustainable aquaculture fisheries and rural development. It accomplished this goal by providing a wide range of technical advice and demonstration projects to governments to indicate how sustainable development and increased production can be achieved consistently with environmental, social, and economic considerations.

FAO has a Department of Fisheries that conducts specific programs focusing on sustainable fishery development for food security purposes with particular emphasis on fishery resources, fishery policy, fishery industries, and fishery information dealing with:

1. Promotion of responsible fisheries sector management at the global, regional, and national levels with priority given to implementation of the FAO-developed Code of Responsible Fisheries (including aquaculture pursuits) and the Compliance Agreement and International Plans of Action relative to the aforementioned Code. FAO is placing special emphasis in the concerns relative to the codes problem dealing with excess harvesting capacity, methods to strengthen regional fisheries bodies, and increasing

FAO's role implementing international instruments to prevent overfishing.

2. Promoting the increased contribution of responsible fisheries and aquaculture to world food supplies and food security. For this objective, the FAO Department of Fisheries is focusing on reducing bycatch (waste or discards) in wild-caught fisheries and promoting responsible aquaculture development in FAO's special program on food security in areas of high potential and critical needs that can be developed consistently with responsible societal and environmental considerations.
3. Global monitoring and strategic analysis of fisheries, with priority given to development of databases and producing various needed detailed analytical reports that contribute to the biennial publication of FAO's "the State of the World Fisheries and Aquaculture." FAO also produces many special reports dealing with aquaculture, such as "Aquaculture and Risk Management" (FAO 1989), "Fisheries and Aquaculture Issues in Small Island Developing States" (FAO 1999), and "FAO Technical Guidelines for Responsible Fisheries—Aquaculture Development" (FAO 1997) that serve as a road map for responsible aquaculture development in areas under national jurisdictions as well as those within transboundary ecosystems. The Code also addresses the responsible use of genetic resources for the purposes of aquaculture, as well as the myriad of other considerations that are necessary for production.

FAO probably maintains the world's largest database and information source accessible via the Internet dealing with fisheries matters (including aquaculture). For example, at their renowned WAICENT Fishery and Aquaculture website, information can be found regarding aquaculture production, aquatic ecology, general aspects of fisheries and aquaculture, fishery production, and links to other associated websites.

The activities of FAO's Department of Fisheries receive oversight and review by the Committee on Fisheries (COFI). This Committee also provides international forums for governments to address major complex fisheries and aquaculture issues and to develop option recommendations for use and resolution by other governmental organizations. The Committee also reviews specific issues relating to fisheries and aquaculture as directed by FAO's Governing Council, the Director-General, or committee member countries or at the request of the United Nations General Assembly. It is important to note that COFI's activities augment or supplement those of other organizational entities working in fisheries and aquaculture, as opposed to superseding them.

Currently, COFI is forming a Subcommittee on Aquaculture to serve as an intergovernmental mechanism for information exchange, discussion, and consensus building on emerging aquaculture issues (including guidance to governments and international bodies on technical and policy matters) (FAO 2001). The formation of the Aquaculture Subcommittee indicates FAO's recognition of its growing contribution to global food security, economic development, and international trade in aquacultured products.

WORLD HEALTH ORGANIZATION OF THE UNITED NATIONS

The WHO of the United Nations is the premier international organization whose mission is to ensure the attainment of the highest level of health by all people. For WHO purposes, health is defined as "a state of complete physical, mental, and societal well-being and not merely the absence of disease or infirmity." WHO was founded in 1948 and has four main functions (1) to provide international guidance in the field of health; (2) to establish global standards for health; (3) to assist national governments in improving their health plans; and (4) to engage in developing and transferring health technologies, standards, and information.

The WHO consists of 191 member countries, administered through a World Health Assembly, Executive Board, and Secretariat located in Geneva, Switzerland. It maintains and provides service to six regional areas: Africa, the Americas, Southeast Asia, Europe, eastern Mediterranean, and western Pacific.

WHO conducts numerous food safety activities, and along with FAO, is a joint sponsor of Codex. The Organization has also sponsored numerous expert consultations and study groups dealing with Risk Assessment, Management, and Communication; Microbiological Standards for Foods; HACCP; Foodborne Trematodes Infections; and Food Safety Issues Associated with Aquaculture and/or Wild-caught species. Of particular significance, two reports focus on aquaculture considerations. They are (1) "Control of Foodborne Trematode Infections" (WHO 1995a) and (2) "Food Safety Issues Associated with Products from Aquaculture." The latter publication clearly indicates that ". . . the production of safe foods from aquaculture was the shared responsibility of governments, industry and consumers, each having an important role to play in the protection of public health . . ." The report further indicates that ". . . food safety concerns associated with aquaculture are predominantly a problem in communities where eating raw or inadequately cooked fish is a cultural practice . . ." and ". . . control strategies should focus on bringing about changes in traditional consumption practices . . ." (WHO 1999).

WHO maintains a global data and information infrastructure dealing with food safety and infant feeding, food legislation and control, a global task force on cholera control, fish food safety, food safety health education, street-vended foods education and training, assessment of food technologies, monitoring chemical contaminants in foods, epidemiological surveillance of foodborne diseases, technical assistance to WHO members, and recognition and collaboration with approximately 200 diverse nongovernmental public health oriented organizations. All of the WHO food safety activities have relevance to aquaculture pursuits.

EMERGING PUBLIC POLICY ISSUES

Within the international trading community for wild-caught species and those derived from aquaculture, numerous public policy issues have been resolved nevertheless numerous others are arising that will need resolution in the future. Properly applied aquaculture technology should provide products that pose no more, or in some cases even fewer hazards, than those associated with traditional wild-caught species. Nevertheless, concerns associated with aquaculture can be complex, and it is important to understand that, as with any technology, there can be shrouded hazards that can have either direct or future impacts and unintended consequences on human, animal, and environmental health issues (Garrett et al. 2000). The globalization of aquaculture endeavors, while multiplying the complexity in achieving broadly based solutions through international deliberations and consumer protection resolutions, can only be obtained through the various formal forums sponsored by the global governmental organizations detailed in this chapter. Such broadly based global forums can, as emerging issues arise, provide the international deliberation platforms and structures necessary to define problems, so that all relevant information can be collected, debated, and then resolved.

Recognizing that aquaculture products are generally used for food and the restocking of wild stocks, aquaculture technology will always be intricately intertwined with emerging controversial issues associated with technological advances in the food and animal husbandry industries. Recognizing this inescapable fact, the aquaculture industry can be expected to become embroiled in highly visible emerging public policy issues. These issues deal with genetically modified foods (GMF), traceability, food safety objectives (FSO), determining the equivalency of inspection services, and how other legitimate factors (OLF) can best be incorporated into the scientific decision making process necessary for regulatory development, as well as the need to ensure traceability in the risk assessment and management decision-making processes. In the author's opinion, it will take many years of debate and consensus building before these emerging public policy issues are resolved.

Genetically Modified Foods

Genetically modified foods (GMF) and the entire discipline of biotechnology is an exciting development that not only offers great opportunities but also presents significant challenges and uncertainties requiring increased public debate on its implications. Biotechnology has widespread applications for foods (including aquaculture), medical innovations, and even wild stock resource restoration (such as fishery resources). However, because perils and differing perceptions almost always accompany progress, it is necessary for these uncertainties to be defined, resolved and accepted by the consuming public so that the technology can progress to its best and fullest utilization. The two principal international organizations dealing with biotechnology and GMF are Codex and the OECD.

Within Codex, the conduct of actions relating to GMF currently resides in a newly constituted group known as the Codex ad hoc Intergovernmental Task Force on Foods Derived from Biotechnology and the Codex Committee on Food Labeling (CCFL).

The new ad hoc Intergovernmental Codex Task Force on Foods Derived from Biotechnology has a terms of reference to develop standards, guidelines, and/or recommendations for biotech-derived foods or traits introduced into foods through the application of biotechnology.

Recent major agreements accomplished in this Codex task force include the development of Draft General Principles for the Application of Risk Analysis (including science-based decision-making, premarket assessment, transparency, and post market monitoring). The task force has also developed Guidelines for Risk Assessment for Food Safety and Nutrition (including "substantial equivalence," potential "long-term health effects," and "nonintentional health effects"), has reviewed the areas of methodology, considered discussion papers on traceability and familiarity, and deferred discussions on the "Precautionary Principle or Approach" and "Other Legitimate Factors" until these concepts mature within other Codex Committees. The task force has also agreed that the primary responsibility for labeling biotech foods rests within the CCFL.

GMF are not a new issue within Codex because its Labeling Committee has been debating the issue since 1994. In terms of labeling, there are several concerns that deal with a definition for GMFs. For example, should they be referred to as foods derived from "modern biotechnology," or "genetic modification," or "genetic engineering"? The task force and the CCFL are not in agreement on this definitional issue.

There is also the issue of mandatory vs. limited mandatory labeling of GMF. Most, if not all, countries agree that if allergens are introduced by biotechnology, then those foods should require mandatory GMF labeling. Alternatively, some other countries want all biotech foods labeled as a "process" consideration, whereas still other countries believe that such mandatory labeling is not required unless the nature of the food has undergone substantial changes relating to its composition, nutritional value, or intended use. This will remain a polarized issue for sometime before resolution is reached.

Traceability

Traceability is another emerging concept that is without a commonly understood definition or method for application and is a confusing concept. It is vital that there should be an agreed-upon definition for traceability within the framework of Codex and agreed-upon General Principles for its application in national risk management schemes. It should be understood that traceability is important in terms of general food safety issues and should be considered broadly. Traceability should take into account existing effective and practical systems based on product coding that allows for the traceback of a manufactured product for possible recall product scenarios. Furthermore, it should also be understood that food products and/or ingredients must be safe before they are placed into the market and, therefore, sufficient food safety assessment and control systems must be in place prior to product distribution to ensure that preventive food safety requirements are met. Product recall and traceback actions, although necessary, should be rare exceptions in effective food control systems and should be used for process failures or other adverse food safety failure scenarios.

One of the difficulties in the international debate on traceability is that because of its ill-defined purpose, definition, and application, it has spawned an international clash of regulatory paradigms. The EU sees traceability, particularly when incorporating identity preservation (IP) programs, as a way to restore consumer confidence relative to bio-engineered foods, dioxin in cattle feeds, and other food safety episodic events that occurred within the EU toward the end of the Twentieth Century. The EU's view is that consumer confidence will be strengthened if foods and ingredients are clearly labeled and can be traced backward to the source and forward to the consumer. In aquaculture lexicon, it is a mandatory "pond-to-plate" regulatory identification program for each fish or farm. The United States' view is that such a detailed identity regulatory imposition extends far beyond that which is necessary for health protection for consumers and will result in excessive product costs and creating impractical regulatory barriers. The United States' view (as the debate is currently framed) is that traceability, as defined by the EU as a regulatory concept, is flawed and fails to deal with problems before they occur. Therefore, the HACCP approach coupled with traceback (where products are identified for recall purposes) as prerequisite manufacturing programs, is a much better food safety approach.

Food Safety Objectives

FSO are an exciting concept that burst upon the scene (primarily Codex) recently during the discussions and agreed-upon global understanding of the HACCP concept. This is an emerging concept for use by either the food producer/processor/buyers or the official agency having jurisdiction. The concept requires that when specific control measures or regulations are initiated, the objective (in this case, a food safety objective) of each specific control procedure be fully documented by a qualitative or quantitative description of what the control procedure is expected to provide for public health protection. When the concept is fully functional and executed, it will provide for benchmark standard measurements for either complete food safety control systems or individual control measures. Part of the difficulty in defining an FSO is determining how broad the application should be. If too broadly applied, it quickly becomes confused with the SPS Acceptable Level of Protection (ALOP) concept. The Codex Committee on Food Hygiene is addressing this issue and is leading the development of the concept within Codex. The concept and application, when fully developed, could be easily modified to meet the needs of OIE, IPPC, and the environmental community and would provide for benchmarking of control activities of national and international systems.

Determining the Equivalency of Inspection Services

Determining the equivalency of inspection services, which is always a difficult task, falls within the purview of the Codex Committee on Food Import and Export Inspection and Certification Systems (CCFICS). It is well recognized that food inspection and certification systems operating in exporting countries may differ from those of importing countries. There are many legitimate reasons for these variations including different hazard profiles, selection of alternative control measures, and national choices concerning the management of food safety risks. It is necessary in these circumstances to determine that the effectiveness of the regulatory exporting food safety system meets the ALOP of the importing country. This has led to the need to develop a protocol to determine export/import food safety systems equivalency, which the Committee is working toward. As one step in that process, the CCFICS is developing Guidelines on the Judgement of Equivalence of Sanitary Measures associated with Food Inspection and Certifications. The Committee is also developing Guidelines for Generic Official Certificate Formats and the Production and Issuance of Certificates, Guidelines for Food Import Control Systems, Guidelines to Judge Equivalence of Technical Regulations Associated with Food Inspection Certification Systems, and Guidelines for Utilization of Industry Quality Assurance that can be used by regulatory food control systems. This Committee's final outputs on the aforementioned items will have a direct effect on the manner in which food will be regulated for food safety purposes nationally and internationally. Similar activities either are under deliberation or have been resolved for the plants and animals that are covered by the terms of reference of the IPPC and OIE, respectively.

Other Legitimate Factors

As indicated above in this chapter, Codex decisions must be premised on sound science to ensure that the organization's standards and recommendations can withstand the most rigorous scientific scrutiny. However, in the organization's statements of Principles Concerning Role of Science in the Codex Decision Making Process and the Extent to Which Other Factors Are Taken into Account, there is a statement that indicates that ". . . when elaborating and deciding upon food standards, the organization will consider, when appropriate, Other Legitimate Factors (OLF) relevant for the health protection of consumers and for the promotion of fair practice in food trade . . ." (FAO/WHO

1997b). The reason for the incorporation of the OLF concept is to give recognition that some legitimate concerns of governments, when establishing their national legislation, may not be globally applicable or relevant. Codex is engaging in the necessary discussions to develop criteria for adoption of the OLF concept. Obviously when developing the OLF criteria, specific necessary actions or safeguards must be incorporated to ensure that the considerations of OLF do not affect the scientific basis of risk analysis conclusions of the organization's recommendations. Furthermore, the interaction of OLF in risk management should not create unjustified barriers to trade.

Transparency

Often corporate or regulatory risk assessments are conducted during the turmoil of dealing with chaotic episodic events to address critical specific public, animal and environmental health issues. This will never change and is akin to being a parent of a teenager. It is vital, however, that when dealing with proactive or reactive risk assessment and management events, the activities and outputs are systematically and fully documented so that any limitations of the process or conclusions are transparent. To ensure transparency, it is absolutely imperative that a formal record be made and be available for review. The record should include any constraints, uncertainties, and assumptions that impacted the risk assessment and management conclusions, as well as any expert judgments that were used in the processes and the rationale for those judgments in the context of the risk scenario. In the author's opinion, this is a fundamental issue that is often not accomplished well by corporate or regulatory officials.

CONCLUSION

Without a doubt, the global trade of fishery products is increasingly complex in terms of (1) the number of countries engaging in the trade, (2) the wide differences in technological sophistication among and between countries, (3) the wide array of different species (more than 500) of fishery products traded, and (4) the multiplicity of regulatory control paradigms that exert significant influence on national, regional, and international fishery import/export control schemes. Experience in the food industry has shown that the food marketplace is changing, and these transformations create new challenges. It has been said that a dinner entree can consist of ingredients from many countries and global regions. For example, tilapia with mango chutney could contain aquacultured fish from Israel as the base, thickeners or gums from Germany, mango puree from Ecuador, ginger from numerous countries in Asia, and spices from India. Institutional food service and home dining events are becoming global melting pots. This has increased the need to rely on food safety systems employed by growers, processors, and various national/international regulatory authorities. Multiplying the complexity of the global fishery trade scenario is the increasing public awareness of the need to assess the effectiveness and improve, where necessary, the societal oversight and regulatory risk management activities associated with all facets dealing with the production, processing, and international trading of foods. Wild-caught fishery and aquaculture endeavors are included in this new public awareness.

As food production technologies advance and more public awareness occurs, relative to the myriad of societal issues that interface with global food production and the international trading of products, aquaculture stakeholders can be expected to be swept into the vortex of the international public debate as a variety of technological production and processing issues are addressed. One might conclude that the intricately intertwined refractory issues dealing with genetically modified foods, traceability, food safety, animal or environmental risk management objectives, determining equivalency of inspection systems, incorporating other legitimate factors into scientific decision-making, and ensuring transparent processes represent a jungle that only Tarzan could untangle. Such is not the case. The international organizations described in this chapter have had remarkable successes in upgrading global human, animal, and environmental health protection, and there is no reason to expect any less as they begin to address the aforementioned and other difficult emerging issues. Although it is true that the contentious issues that Codex, OIE, IPPC, and ISO are now finding themselves debating have become more difficult because of the new world stature and global trade importance that WTO has conveyed to each organization, each organization is up to the task. However, longer consensus time lines in resolving controversial issues can be expected, as protracted debates and more transparent processes occur. In the author's opinion, among the animal protein groups, fishery products probably have had the greatest number of different country and regional interactions in terms of global import/export trade considerations. In many respects, because of that global trading history, the fishery products industry and those who regulate it are leading the way in innovative international trade regulation and development.

The role for aquaculture stakeholders in the global discussions on increased regulation is to become more aware of the international organizations in which these discussions take place, their detailed processes, and take part in the global debates. Aquaculture stakeholders must acknowledge that although perils often accompany progress, properly applied aquaculture technology should provide products that pose no more, and in certain instances, even fewer hazards than some associated with more traditional wild-capture species. Recognizing in some instances that all perils are not completely understood by new or modified global technologies, an integrated public, animal, and environmental risk assessment, management, and communication approach is necessary to minimize unintended consequences from developing technologies. The challenge for aquaculture stakeholders is to require that the technologies associated with their interests are developed and applied in an integrated approach that builds on the totality of issues and past experiences, to ensure that aquaculture neither causes nor gives the appearance of contributing to unacceptable risks that could negate the improved economic, nutritional, and resource enhancement benefits that aquaculture technology offers.

REFERENCES

Badiru, A. 1995. Industry's Guide to ISO 9000. John Wiley and Sons, Inc., New York, NY.

Burke, R. 2000. What is this thing called Codex and why is it so important? Presented at AFDO International Workshop, Burlington, VT, June 17.

Crosby, P.B. 1984. *Quality Without Tears: The Art of Hassle-Free Management.* McGraw-Hill, New York, NY.

Darby, W.J., P. Ghalioungui, and L. Grivetti. 1977. *Food: The Gift of Osiris.* Academic Press Inc., London. p. 338.

Dillon, M. and C. Griffith. 1997. Standards and specifications. *In*: How to Audit, pp. 16–29. M.D. Associates, Grimsby, UK.

FAO. 2001. FAO's Committee on Fisheries Establishes New Subcommittee on Aquaculture: One of the fastest growing food production sectors. Press release 01/12. Food and Agriculture Organization of the United Nations, Rome, Italy.

FAO. 2000. Commodity Market Review 1999–2000. Food and Agriculture Organization of the United Nations, Rome, Italy. pp. 85–89.

FAO. 1999. Fisheries and Aquaculture Issues in Small Island Developing States COFI/99/7. Considered at the 23rd Session of the Committee on Fisheries, Rome, Italy.

FAO. 1997. IMO/FAO/UNESCO-I0C/WMO/WHO/IAEA/UN/UNEP Joint Group of Experts on the Scientific Aspects of Marine Environmental Protection (GESMAP) towards safe and effective use of chemicals in coastal aquaculture. Food and Agriculture Organization of the United Nations, Rome, Italy.

FAO. 1995. FAO Code of Conduct for Responsible Fisheries—Aquaculture Development. Food and Agriculture Organization of the United Nations, Rome, Italy.

FAO. 1989. ADCP/REP/89/41—Aquaculture and Risk Management. Food and Agriculture Organization of the United Nations, Rome, Italy.

FAO/WHO. 1999a. Origins of the Codex Alimentarius. *In*: Understanding the Codex Alimentarius. Food and Agriculture Organization of the United Nations, Rome, Italy.

FAO/WHO. 1999b. Draft Principles and Guidelines for the Conduct of Microbiological Risk Assessment. CAC/GL-30 (1999). Food and Agriculture Organization of the United Nations, Rome, Italy.

FAO/WHO. 1998. Guidance on Regulatory Assessment of HACCP. Report of a Joint FAO/WHO Consultation on the Role of Government Agencies in Assessing HACCP. World Health Organization, Rome, Italy.

FAO/WHO. 1997a. Codex Alimentarius Procedural Manual, 10th ed. Food and Agriculture Organization of the United Nations, Rome, Italy. pp. 69–72.

FAO/WHO. 1997b. Codex Alimentarius Procedural Manual, 10th ed. Food and Agriculture Organization of the United Nations, Rome, Italy. pp. 146.

FAO/WHO. 1995. Application of Risk Analysis to Food Standards Issues. Report of the Joint FAO/WHO Expert Consultation. World Health Organization, Geneva, Switzerland.

Garrett, E.S. 2001a. Emerging Food Safety Considerations in the Global Market Place. Presented at the Fourth Biennial Conference on Fish Processing, Grimsby, UK, July 3.

Garrett, E.S. 2001b. Along the yellow brick road towards risk assessment . . . what to trace? Presented at Building Effective Traceability Systems, The Town Hall, Grimsby, UK, July 6.

Garrett, E.S., M.L. Jahncke, and E.A. Cole. 1998. Effects of Codex and GATT. J. Food Control 60(2–3): 172–182.

Garrett, E.S., M. Jahncke, and J. Tennyson. 1997. Microbiological hazards and emerging food safety issues associated with seafoods. J. Food Protection 60(11): 1409–1415.

Garrett, E.S., M. Hudak-Roos, and M. Jahncke. 1994. The international trading of seafoods—issues and opportunities. Presented at the International Food Technologist Blocks in Trading Blocks and Regulatory Opportunities, Atlanta, GA, June 27.

Garrett, E.S. and M. Hudak-Roos. 1991. Developing an HACCP-based inspection system for the seafood industry. J. Food Technol. Dec. pp. 53–57.

GATT. 1991. WTO archives. United States—Restriction on Imports of Tuna. Report not adopted, circulated on 3 September 1991. Geneva, Switzerland.

GATT. 1988. WTO archives. Conciliation. Canada—Measures Affecting Exports of Unprocessed Herring and Salmon. L/6268-35S/98—Report of the Panel adopted on 22 March 1988. Geneva, Switzerland.

GATT. 1982. WTO archives. Conciliation. United States—Prohibition of Tuna and Tuna Products from Canada. L/S5198-295/91—Report of the Panel adopted on 22 February 1982. Geneva, Switzerland.

Herborg, L. 1997. HACCP-Based Inspection Programs. *In*: Fish Inspection, Quality Control and HACCP (eds. R. Martin, R. Collette, and J. Slavin), pp. 679–686. Technomic Publishing Company, Inc., Lancaster, PA.

ISO. 2001. Introduction to ISO. International Standards Organization, Geneva, Switzerland.

ISO. 1996. ISO 14000 Case Studies Models for Implementation. Ceem Information Services, Fairfax, VA.

Llaneras, M. 2000. Seaweed products: brown algae of economic significance. *In*: Marine and Freshwater Products Handbook (eds. R.E. Martin, E.P. Carter, G.J. Flick, and L.M. Davis), pp. 531–540. Technomic Publishing Company, Inc., Lancaster, PA.

Lovell, R. 2000. Aquaculture. *In*: Marine and Freshwater Products Handbook (eds. R.E. Martin, E.P. Carter, G.J. Flick, and L.M. Davis), pp. 847–857. Technomic Publishing Company, Inc., Lancaster, PA.

National Marine Fisheries Service (NMFS). 2000. Fisheries of the U.S. Current Fisheries Statistics. 1999. U.S. Department of Commerce, NOAA, NMFS. Silver Spring, MD.

OECD. 1996. Reconciling Pressures on the Coastal Zone—Fisheries and Aquaculture. Organization for Economic Co-operation and Development, Paris, France.

OIE. 2000. 68th Annual General Session of the International Committee of the Office International des Epizooties. Press release, 26 May. World Organization for Animal Health, Paris, France.

OIE. 2001. International Aquatic Animal Health Code—2001. World Organization for Animal Health, Paris, France.

Peach, R.W. 1992. The ISO 9000 Handbook. Ceem Information Services, Fairfax, VA.

Rainosek, A. 1997. Inherent risks in acceptance sampling. *In*: Fish Inspection, Quality Control and HACCP (eds. R. Martin, R. Collette, and J. Slavin), pp. 530–537. Technomic Publishing Co., Inc., Lancaster, PA.

Regenstein, J.M. and C.E. Regenstein. 2000. Religious food laws and the seafood industry. *In*: Marine and Freshwater Products Handbook (eds. R.E. Martin, E.P. Carter, G.J. Flick, and L.M. Davis), pp. 477–486. Technomic Publishing Co., Inc., Lancaster, PA.

Rudolph, B. 2000. Seaweed products: red algae of economic significance. *In*: Marine and Freshwater Products Handbook (eds. R.E. Martin, E.P. Carter, G.J. Flick, and L.M. Davis), pp. 515–529. Technomic Publishing Co., Inc., Lancaster, PA.

United States Trade Representative (USTR). 1994. Final Act Embodying the Uruguay Round of Multilateral Trade Negotiations. *In*: Uruguay Round of Multilateral Trade Negotiations General Agreement on Tariffs and Trade, p. 7. US Govt. Printing Office, Washington, D.C.

WHO. 1999. Food Safety Issues Associated with Products From Aquaculture—Report of a Joint FAO/NACA/WHO Study Group. World Health Organization, Geneva, Switzerland.

WHO. 1996. Guidelines for strengthening a National Food Safety Programme. World Health Organization. Geneva, Switzerland.

WHO. 1995a. Control of Foodborne Trematode Infections—Report of a WHO Study Group. World Health Organization, Geneva, Switzerland.

WHO. 1995b. Training aspects of the Hazard Analysis Critical Control Point System (HACCP). Report of a WHO Workshop on Training in HACCP with the Participation of FAO. World Health Organization, Geneva, Switzerland.

WTO. 1998a. WT/DS18/R—Australia—Measures Affecting Importation of Salmon Report of the Panel. December. pp. 207–221. Geneva, Switzerland.

WTO. 1998b. United States—Import Prohibition of Certain Shrimp and Shrimp Products—Report of the Appellate Body (AB-1998-4) Geneva, Switzerland.

WTO. 1995a. Results of the Uruguay Round of Multilateral Trade Negotiations, 1993. Agreement on Application of Sanitary and Phytosanitary Measures. December. Geneva, Switzerland.

WTO. 1995b. Results of the Uruguay Round of Multilateral Trade Negotiations, 1993. Agreement on Technical Barriers to Trade. December. Geneva, Switzerland.

6

Future Considerations of Global Aquaculture

Michael L. Jahncke

INTRODUCTION

Traditional capture fisheries will be unable to supply the increasing global demand for fishery products. Future supplies of such products will be dependent on aquaculture, which has grown on a worldwide basis at an annual rate of approximately 11% since 1984, compared with 3% and 1.6% for livestock meat and capture fisheries, respectively (FAO 1997). Freshwater aquaculture focusing on herbivorous/omnivorous species (e.g., carps) will continue to be the primary area of production of low-cost species, for developing countries to meet their increasing food requirements. Additionally, coastal and marine aquaculture operations will face increased environmental and societal pressure, which will require production methods that are more environmentally friendly (FAO 1997).

Aquaculture issues are complex and dynamic; however, there are risks associated with any agricultural production industry. Goldberg et al (2001) identified several risks to the environment from marine aquacultures. Sustainable aquaculture is the future of the industry. Sustainability can be defined as "an integrated system of plant and animal

Public, Animal, and Environmental Aquaculture Health Issues,
Edited by Michael L. Jahncke, E. Spencer Garrett, Alan Reilly,
Roy E. Martin, and Emille Cole.
ISBN 0-471-38772-X (cloth)

production practices having a site-specific application that will over the long term: (1) satisfy human food and fiber needs; (2) enhance the environmental quality and natural resource base upon which the agricultural economy depends; (3) make the most efficient use of nonrenewable resources and on-farm resources and integrate, where appropriate, natural biological cycles and controls; (4) sustain the economic viability of farm operations; and (5) enhance the quality of life for farmers and society as a whole" (USDA 1977).

In 1971, Philip Roedel (former director, US Bureau of Commercial Fisheries, which is currently the US National Marine Fisheries Service) identified 10 important issues that needed to be addressed to continue the growth of freshwater and marine aquaculture. It is interesting to note that most of these issues are still relevant today.

AQUACULTURE ISSUES 1971–2000

Roedel (1971) stated that the year 2000 "may be considered a focal point for projecting the future of the US fisheries." He was correct. However, today we realize that he should have been broader in his scope to include the world's fisheries. Roedel also stated, "To meet our increasing demand for certain fishery products, we must look still further. Artificial propagation of freshwater, brackish-water, and marine animals offer considerable promise for the future." This declaration is still true today.

Roedel (1971) identified 10 issues that needed to be addressed to ensure the future success of aquaculture. These issues were:

1. More knowledge on the physiological and ecological requirements of culture animals, diseases and their prevention, and control of predators and competitors;
2. Development of selectively bred strains with desirable characteristics;
3. Improvement in the design of ponds, enclosures, and hatcheries for optimum growth and survival;
4. Engineering studies to develop new approaches to harvesting and handling of fish;
5. Studies to determine the relationship between the environment in which the organisms are cultured and the quality of the product;
6. Development of techniques for disposing of by-products from processing operations;

7. Protection of the environment from adverse changes and plans for rational use of coastal and high seas waters;
8. Government support of research, development, and establishment of management systems that permit effective conservation of the resource;
9. Development of latent fishery resources and providing the consumer with acceptable food in terms of safety, quality, and cost; and
10. Long-range planning for fishery management purposes.

He also said that the future of aquaculture was dependent on the combined efforts of biologists and engineers to develop techniques for the mass culture of organisms, improve harvest methods, increase available capital for large scale ventures, and have a mechanism in place to help resolve environmental conflicts among user groups (Roedel 1971).

AQUACULTURE ISSUES 2001–2020

The 1997 FAO Report on the State of World Aquaculture listed six indicators that pointed to the continued growth of global aquaculture.

The six indicators as stated in the report are:

1. **Increasing demand:** "Increasing demand for fish will require more aquaculture production, because the supply from capture fisheries is static."
2. **Emergence of the sector:** "Aquaculture has become recognized as a growth sector of economic importance in many countries and has attracted the attention of the private and public sectors. Development plans of most producing countries are aimed at increasing fish supplies from aquaculture for local and export markets and at increasing the sector's contribution to food security in rural areas."
3. **Vertical expansion of production:** "In many developing countries, including major producers, there is: (i) potential for fairly simple and relatively inexpensive means to increase production in many existing systems; (ii) an evident trend toward sustainable intensification coupled with the existence of infrastructure used for extensive traditional production, especially in Asia; and (iii) good potential for increasing productivity (e.g., through more efficient management and improved management skills, better feeds and

feeding strategies, reduction of loss to disease, genetic improvement of stocks)."

4. **Horizontal expansion of production:** "Production areas are being expanded in China, India, and other main areas, by mobilizing land resources unsuited to agriculture (saline soils, water clogged areas, etc.), underutilized water bodies (for cage and pen culture), seasonal water bodies, and rain-fed ponds. Expansion is also occurring by use of irrigation systems and by integration of aquaculture with agriculture. In the long term, integration of aquaculture into watershed management and coastal zone management will provide growth opportunities by facilitating competitive access to resources. Horizontal expansion, although modest in the near term, will also accrue from new producing countries."
5. **Culture-based fisheries:** "Stocking of reservoirs and enhancement/rehabilitation of fisheries will gain importance with time, particularly as cost/benefit problems are resolved."
6. **Growing awareness of sustainability needs:** "There is a rapidly growing awareness of the need to ensure the sustainability of the sector in the long term. Public debate involving all stakeholders, national and international efforts to arrive at practical guidelines for sustainable practices (codes of practice), and technical efforts to improve the sustainability of some aquaculture systems, are positive responses to challenges and will yield constructive results in the medium- and long term."

Development prospects will vary considerably among countries and regions depending on a number of variables, discussed below. Translating the outlook for development into projected production levels is not a simple matter. "Several key variables intervene in defining outlook at the national level, whereas the global outlook could be interpreted as the sum of national outlooks, modified by synergisms and antagonisms resulting from interactions between countries" (FAO 1997).

SUSTAINABLE AQUACULTURE

Sustainability is the overriding strategic issue and challenge to all economic sectors, including aquaculture (FAO 1997). The aforementioned FAO report identified the following six issues needed for sustainable aquaculture:

1. **Establishment of an enabling environment:** "Strengthen institutional capabilities, decentralized management, new institutional arrangements/appropriate institutional linkages, and new legal and administrative frameworks. Policies and plans are also needed that include more funding, improved communications and collaboration among stakeholders, increased allocation of resources to aquaculture, zoning issues, and greater public support for aquaculture."
2. **Adoption of an integrated planning and management framework:** "Integrate aquaculture into catchment and coastal area management programs. As a managed resource, it then becomes part of an integrated rural and agricultural development program."
3. **Increasing the efficiency of resource use and productivity in general at the farm level:** "Adoption of a systems approach to aquaculture management, lowering water system requirements/improved water management practices, better feeding strategies, and less polluting feeds. In addition, genetically improved stocks, improved health management programs, and integration of aquaculture with agriculture are needed."
4. **Reducing externalities and avoiding irreversible damage:** "Avoid negative human and environmental impacts of aquaculture through consultative planning, risk assessment, improved systems management programs, and improved site selection and zoning criteria. Implement preventive health management/vaccination programs and design better-engineered farms to reduce effluents and improve waste treatment procedures."
5. **Establishing reliable databases and effective information management systems:** "Increase the awareness of all stakeholders and the general public on the need for the sustainability of aquaculture."
6. **Maximizing positive sustainable attributes:** "(i) Food security; (ii) conserve genetic material in support of maintenance of biodiversity through stock rehabilitation and cryopreservation of gametes; (iii) reduce pressure on key hunted stocks by providing farmed alternatives; (iv) improve coastal waters through the culture of mollusks and aquatic plants; (v) develop small-scale integrated aquaculture-agriculture farming systems; and (vi) complement restoration efforts and contributions to pest control (FAO 1997)."

In addition to those items identified under the six broad categories in the FAO report, other issues include water supply, aquaculture site selection criteria, escapements, effluents, the continued use of fish meal in formulated aquaculture diets, and transgenics. The latter two emerging topics will be discussed in this chapter. Several of the other issues are beyond the scope and purpose of this book, and others such as fishery stock issues, public, animal, and environmental health issues, escapements, effluents, HACCP, and trade issues were discussed in Chapters 1 through 5 and will be summarized in this chapter.

FISHERY STOCK ISSUES

Marine harvest of wild stocks from most of the primary fishing areas in the Atlantic Ocean and some areas of the Pacific Ocean are at their maximum sustainable yields (FAO 2000). For the world as a whole, landings of marine fish have leveled off, and sustainable aquaculture is needed to help meet and supplement some of the world's food protein requirements. The FAO definition of aquaculture is "the farming of aquatic organisms, including fish, molluscs, crustaceans, and aquatic plants." Farming implies human intervention in the rearing process to enhance production practices such as stocking, feeding, health maintenance, and predator protection (FAO 1995). Farming also implies individual or corporate ownership of the stock under cultivation.

In 1999, 34% of fishery production on a worldwide basis was farm raised, with approximately 90% of all aquaculture being produced in Asia. Nonindustrialized countries accounted for approximately 87% of all aquaculture production. By 2025, the world's population is expected to be approximately 8 billion people, with nonindustrialized countries experiencing the largest increases in population growth. In these countries, along with the continued emphasis on the culture of fast-growing species to feed their populations, efforts are being directed to develop intensive culture methods for high-valued species to increase both exports and revenues. Industrialized countries, on the other hand, need to develop sustainable aquaculture industries to decrease their reliance on fisheries imports.

PUBLIC HEALTH ISSUES

Public health issues of aquaculture animal products in both nonindustrialized and industrialized countries focus on biological hazards (e.g., parasites, bacteria, and viruses) and on chemical hazards (e.g., chemicals, pesticides, chemotherapeutants, heavy metals) (WHO 1999). Parasite issues in industrialized countries are definite but are not considered a major public health issue and are primarily associated with the consumption of raw or undercooked products. However, fishborne trematodiasis (e.g., *Clonorchis* and *Opisthorchis)* is a serious disease in many parts of the nonindustrialized world, particularly in Asia. Parasite issues in marine aquaculture can be addressed by feeding only commercially prepared dry diets to aquacultured species to prevent infestation of the fish by parasites via their diet. This approach is suitable for species that are raised in net pens or closed recirculating systems, where there is more control over the food fed to the organisms. However, in freshwater aquaculture operations, parasites enter the fish by penetrating the skin. Thus, in freshwater aquaculture ponds, HACCP principles have been successfully used to control trematode population, resulting in trematode-free fish (Lima dos Santos 1994).

In addition to parasites, several species of pathogenic bacteria are indigenous to freshwater and marine environments. Control of pathogens in the aquaculture environment is dependent on the development of management protocols, sanitation protocols, and employee training programs. The available data indicate that there is a low incidence of enteric pathogens in aquaculture systems. However, this may not be true for aquaculture systems that use untreated manure to fertilize the culture system. The available epidemiological data indicate higher incidences of diarrhoeal illnesses in communities consuming fish from ponds fertilized with waste water and manure (Blum and Feachem 1985). Management and regulatory programs are needed to control such bacteria and viral hazards by eliminating the use of human excrement and untreated animal wastes in culture tanks and ponds.

Although chemical hazards (e.g., pesticides, heavy metals, chemotherapeutants, biotoxins) are a worldwide concern, available data indicate that the majority of aquacultured products contain low concentrations of heavy metals and chlorinated hydrocarbons. In fact, aquaculture public health risks from chemical hazards can be minimized by taking a holistic approach that includes (1) proper siting of facilities, (2) monitoring of growing waters (e.g., monitor heavy metals, chemicals, biotoxins), (3) implementation of codes of practice, good agricultural practices (GAPs), best management practices (BMPs), and good manufacturing practices (GMPs) to control chemotherapeutant issues, (4) end-product surveillance testing of products, and (5) training programs for aquaculture employees.

ANIMAL HEALTH ISSUES

Pathogens are a significant problem in both freshwater and marine environments, and they affect both aquacultured and wild species. Before the introduction of aquacultured species, a risk assessment should be conducted. Parasite control includes using a combination of chemical and biological approaches in addition to, quarantine, and early diagnosis. Bacterial and viral diseases can be spread vertically (e.g., parents to offspring via sex cells) and/or horizontally (e.g., one organism to another via direct contact, air, or water). Disease control strategies include implementing BMPs and GMPs, proper administration of drugs under a veterinarian's direction, development of new vaccines, production of high-health animals, and implementation of codes of practice and following good facility management protocols.

ENVIRONMENTAL HEALTH ISSUES

Central planning efforts are required to properly identify appropriate sites for aquaculture operations to ensure that they are both environmentally and socially friendly. The FAO Code of Conduct for Responsible Fisheries is global in its approach and addresses issues from production to the consumer (FAO 1995). Many countries are currently in the process of developing codes of practice for their aquaculture industries to address issues such as escapement, habitat destruction, and effluents. Combinations of water reuse technologies, development of more nutritionally complete diets, and use of settling ponds, artificial wetlands, open ocean siting of net pens, etc. will help to minimize impacts from aquaculture. In addition, the development of new vaccines, the use of high-health organisms, proactive disease management programs, employee training programs, implementation of HACCP principles and BMPs will reduce the levels of pathogenic organisms, chemicals, and chemotherapeutants transferred into the surrounding environment from aquaculture.

HACCP

HACCP is accepted on a worldwide basis as a systematic risk management control program for food safety issues. HACCP food safety programs from farm (e.g., harvest, process, and distribution) to table (e.g., food acquisition, food handling, food preparation, serving, and storage of leftovers) can help control biological, chemical, and physical hazards found in the product. However, in addition to aquaculture food safety issues, HACCP principles can be applied to control animal and environmental issues (e.g., diseases, escapements, effluents, chemicals, chemotherapeutants). As previously noted, HACCP principles have been successfully used in Asia to prevent infestation of pond-cultured fish with trematodes. In the United States, HACCP principles are being used to control exotic pathogens at a vertically integrated shrimp production facility and to help prevent the introduction of exotic pathogens into the environment from shrimp processing operations (Jahncke et al. 2001). HACCP principles, in combination with Codes of Practice, BMPs, and employee training programs, are the basis for sustainable aquaculture.

TRADE OF INTERNATIONAL PRODUCTS

The international trade in fishery products is massive and complex. No other animal protein commodity is as actively traded between countries. The role of the WHO/FAO Food Standards Program has become increasingly important since the Uruguay Round of GATT. Adherence to Codex Standards can offer safe harbor in WTO disputes. The Office International des Epizootics (OIE) is the WHO's program for animal health and is the second of three international health organizations that promulgate standards, which can provide a legal safe harbor in cases of WTO trade disputes. In addition, the mission of the OIE is to inform governments of the occurrence and course of global animal diseases and to identify methods that can be implemented to control such diseases.

Since aquaculture products are used for food purposes, there are two WTO agreements of specific significance to aquaculture: (1) Sanitary and Phytosanitary (SPS) and (2) Technical Barriers to Trade (TBT) (Garrett et al. 1998). The SPS Agreement provides for the basic rules for food safety and animal and plant health standards. The SPS Agreement requirements protect public or animal life from risks due to contaminants, toxins, food additives, or disease-causing organisms. The SPS text preserves a country's ability to maintain its own standards if they are based on science or if they achieve the level of protection that the importing country deems appropriate. The TBT has activities dealing with standards of development and conformity assessment as well as a number of specific considerations that deal with international and regional system requirements. The TBT agreements cover regulations, voluntary standards, and associated conformity assessment procedures used to determine compliance, except when such are specific sanitary or phytosanitary measures defined by the SPS agreement.

The specific mission of the Codex Alimentarius Programme is to protect consumers and to facilitate trade. The Codex Programme has upgraded global food manufacturing practices that have dramatically improved global consumer protection. Codex has a formal mandate governing the role of science for all its decision-making procedures. The Codex approach to HACCP has been well received in terms of the seven general HACCP Principles and Guidelines, logic sequence, use and misuse of decision trees, worksheets, training requirements and application through specific Codes of Practice or GMPs. Some countries question whether international trading partners can mandate HACCP programs for food imports and exports, especially if traditional non-HACCP food control approaches can provide equivalent levels of protection (Garrett et al. 1998).

FISH MEAL

Is the use of wild fishery stocks in the manufacture of aquaculture diets a wise use of a resource? Specifically, should the resource be used directly to feed people, or left for wild stocks to eat? Over the past several decades considerable research efforts have been dedicated to developing human food uses from many industrial species of fish. Efforts have included the production of fish protein concentrate in the 1960s and 1970s (Roedel 1971), conversion of industrial fish into surimi and other analog products (Zapata Haynie Corp. 1989), and the use of these species as intact or minced products. To date, these efforts have had limited success. It is unlikely that future research efforts will be successful in producing high quality food products from these species, because most of the species (e.g., menhaden, anchovies, caplin) are small, bony fish with high oil content and contain fishy flavors that limit their shelf life and suitability for human food purposes. Nevertheless, the ethical question still remains concerning the use of these resources in animal feeds rather than as a direct food for people (Best 1996; Goldberg and Triplett 1997; Hansen 1996; Pimenthal et al. 1996).

Whether or not species should be left as a food resource for other wild fish stocks is an important fishery resource issue. However, because most of these species are currently under state or governmental fishery management programs, and there is considerable information on the life cycles and life histories of these species, the sustainability of these fisheries is controlled for the foreseeable future. The landings for fish meal production have not changed significantly in decades (Hardy 2001). Nevertheless, it is extremely important to reduce the amount of fish meal in aquaculture diets.

Aquaculture is not the world's largest user of fish meal; it is also used in the diets of other animal production industries such as poultry and swine (Naylor et al. 2001). However, the use of fish meal in diets is increasing, as new high-valued species are grown (e.g., shrimp, flatfish, snappers) that require high-quality protein diets. Fish meal is one of the most expensive components of aquaculture diets, and the future viability of aquaculture is based on reducing the cost associated with these diets. Nonindustrialized countries need to seek alternative feed ingredients to reduce their dependence on expensive imported feed ingredients (Best 1996). Thus the high cost of fish meal-based diets is one of the driving forces behind continued research in developing more nutritionally complete diets based on vegetable proteins. Nutritionally complete, vegetable-based protein diets are needed to reduce the cost associated with the production of such high-valued species. Such diets are being developed and will dramatically reduce dependence on fish meal. In addition, high-valued transgenic species are being developed that are capable of using vegetable-based protein diets.

GENETICALLY MODIFIED ORGANISMS

An important emerging issue in aquaculture is the use of genetically modified organisms (GMO). The FAO stated that GMO issues concerning human and environmental health safety issues must be addressed. GMO describe the application of rDNA to the genetic alteration of microorganisms, plants, and animals (Anon. 2000). Potential benefits include increased resistance to disease, adaptability to harsh growing conditions, increased tolerance to chemicals, faster growth, and better sensory characteristics.

The Institute of Food Technologists' Expert Panel on Food Safety concluded that biotechnology has a long history of use in food production, representing a documentable continuum encompassing traditional breeding techniques and new techniques based on molecular modification of organisms. The panel concluded that the use of rDNA biotechnology and molecular methods to genetically manipulate organisms expands the scope of genetic changes that can be made to organisms and broadens the scope of possible food types but inherently does not result in foods that are less safe than those developed by more conventional methods (IFT 2000). However, unintended effects of GMO are speculated to occur by expression of an unknown or unexpected toxic or antinutritional factor or some other enhanced production of known toxicity (Royal Society 1998).

Food safety issues include potential biological activity, allergenicity, and possible toxicity of GMO (Hallerman 2000). Environmental safety issues focus mainly on potential loss of genetic diversity in wild species by interbreeding and displacement of wild stocks from their habitats by escaped cultured GMO. To date, only a few environmental risk assessments have been conducted on the impact of GMO on the environment. These assessments have indicated some potential risk to native species from GMO. Studies have yet to be conducted on transgenic shellfish (Hallerman 2000).

Currently, there is tremendous worldwide consumer resistance to GMO products. Consumer resistance to these products is unlikely to disappear in the foreseeable future. Eighty-five percent of consumers favor explicit labeling of GMO products (Hallerman 2000). Effective risk communication is essential to ensure that the public eventually accepts these products. At the very least, these products must demonstrate that they pose minimal risk to the public, other animals, and the environment. In addition, these products must also be perceived by consumers to have unique and desirable qualities such as improved nutritional quality, improved sensory characteristics, or lower cost (Hallerman 2000).

Industry must work closely with regulatory agencies to ensure that all potential public, animal, and environmental health issues are identified. There are potential benefits from transgenic species; however, oversight by regulatory agencies is necessary and must be based on sound science to safely advance the technology (Entis 2000). It is critical that data be provided to regulatory agencies to demonstrate that these GMO species pose no public health concerns and are nutritionally and chemically comparable to wild species (Morgan 1999).

Methodologies must be developed to produce organisms that are incapable of reproducing if they escape into the wild. Environmental organizations are concerned that escaped transgenic species may crossbreed with wild species resulting in inferior offspring, reducing genetic diversity, and reducing the total numbers of wild stocks (Entis 2000).

PROJECTIONS

A major force pushing the growth of aquaculture has been the declining supplies of wild stocks of many species. Another factor has been the desire for year-round supplies of certain species. By 2025, the world population is expected to reach nearly 8 billion people, with nonindustrialized countries experiencing the largest increases in population growth. These countries must develop aquaculture industries that are capable of meeting their increasing food demands and widening their export earnings. Intensified production methods will continue to increase yields. Several countries have room for incredible growth because of abundant resources, but government and external assistance will be needed to meet these expectations. The aquaculture industries of Europe and North America, on the other hand, will have their growth restrained by government regulations and competing uses for limited resources. Nevertheless, with recent advances in bioengineering, aquaculture producers of the future will have methods available to them that can radically change production practices. A future question yet to be answered is, Will consumers accept transgenic fish? Nevertheless, demand for food will keep aquaculture on the crest of the wave.

REFERENCES

Anon. 2000. Genetically modified organisms (GMO): A backgrounder by the Institute of Food Technologists. J. Food Technol. 54(1): 42–45.

Best, P. 1996. Focus or feed: should fish meal be covered by the ethics of environmentalism? Feed Intl. 17(8): 4.

Blum, D. and R.G. Feachem. 1985. *Health aspects of nightsoil and sludge use in agriculture and aquaculture. Part III. An epidemiological perspective.* Dübendorf, Switzerland: International Reference Centre for Waste Disposal.

Entis, E. 2000. Issues in the commercialization of transgenic fish. Proceedings of the Third International Conference on Recirculating Aquaculture (eds. G. Libey and M. Timmons, G. Flick and T. Rakestraw), Virginia Polytechnic Institute and State University, Roanoke, VA, July 20–23, 2000, pp. 148–155.

FAO. 2000. The state of the world fisheries and aquaculture. ISBN 92-5-104492-9. Food and Agriculture Organization of the United Nations. Rome, Italy.

FAO. 1997. Review of the state of the world aquaculture. FAO Fisheries Circular. No. 886, FIRI/C886 (Rev. 1). ISBN 0429-9329. Food and Agriculture Organization of the United Nations. Fisheries Development, Rome, Italy.

FAO. 1995. Code of conduct for responsible fisheries. Food and Agriculture Organization of the United Nations. Fisheries Department, Rome, Italy. p. 41.

Garrett, E.S., M.L. Jahncke, and E.A. Cole. 1998. Effects of Codex and GATT. Food Control 9(2–3): 177–182.

Goldberg, R. and T. Triplett. 1997. Murky waters: Environmental effects of aquaculture in the United States. Environmental Defense Fund. New York, NY.

Goldberg, R.S., M.S. Elliot, and R.L. Naylor. 2001. Marine Aquaculture in the United States: Environmental Impacts and Policy Options. Pew Oceans Commission. Arlington, Virginia. 33 pp.

Hallerman, E.M. 2000. Commercialization of Modified Fish and Shellfish Policy Issues. Global Aquaculture Advocate Dec. 2000. 3(6): 86–88.

Hansen, P. 1996. Food uses of small pelagics. INFO Fish Intl. 4: 46–52.

Hardy, R.A. 2001. True impact of aquaculture on wild fish. Global Aquaculture Advocate. June 2001 4(3): 6, 8.

Institute of Food Technologists (IFT). 2000. IFT experts report on biotechnology and foods. Human food safety evaluation of rDNA biotechnology-derived foods. Food Technol. Nov. 2000. 54(9): 53–61.

Jahncke, M.L., C.L. Browdy, M.H. Schwarz, A. Segars, J.L. Silva, D.C. Smith, and A. Stokes. 2001. Preliminary application of hazard analysis critical control point (HACCP) principles as a risk management tool to control exotic viruses at shrimp production and processing facilities. *In*: The New Wave, Proceedings of the Special Session on Sustainable Shrimp Culture, Aquaculture 2001 (eds. C.L. Browdy and D.E. Dory), pp. 279–284. The World Aquaculture Society, Baton Rouge, LA.

Lima dos Santos, C.A. 1994. The possible use of HACCP in the prevention and control of foodborne trematode infections in aquacultured fish. Paper presented at the Symposium on New Developments in Seafood Science and Technology, 37th Ann. Conf. of the Can. Institute of Food Sci. and Technol. Vancouver, Canada, May 15–18, 1994.

Morgan, S. 1999. Aqua Bounty Farm uses biotechnology to produce fast-growing fish. Aquaculture Gazette Oct. 1999. 1(5): 1.

Naylor, R.L., R.G. Godbert, J. Primavera, N. Kautsky, C.M. Malcolm, M. Beveridge, J. Clay, C. Fulke, J. Lubchenco, H. Monney, and M. Troell. 2001. Effects of aquaculture on World Fish Supplies. Issues in Ecology Ecological Society of America, Washington, DC.

Pimentel, D., R.E. Shanks, and R.C. Rylander. 1996. Bioethics of fish production Energy and the environment. J. Aquaculture Environ. Ethics 9(2): 144–164.

Roedel, P. 1971. Fish for 300 million. *In*: Our Changing Fisheries (ed. S. Shapiro), pp. 506–521. USDOC, US Government Printing Office, Washington, DC.

Royal Society. 1998. Genetically modified plants for food use. sss.royalsoc.ac.uk/files/statfiles/document-56, pdf.

USDA. 1977. Sustainable agriculture. Natural Agricultural Research, Extension of Teaching Policy Act of 1977. Section 1404 as amended by Section 1603 of the FACT Act.

WHO. 1999. Food safety issues associated with products from aquaculture. Report of a Joint FAO/NACA/WHO study group. WHO Technical Report Series 883. WHO, Geneva, 1999, p. 55.

Zapata Haynie Corporation. 1989. Surimi production for Atlantic menhaden. Final Project Report. May 11, 1989. Contract #50-EANF-6-00048. Zapata Haynie Corp. Reedville, VA.

Index